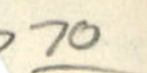

MUSIC, SOUND AND SENSATION

MUSIC, SOUND AND SENSATION

A MODERN EXPOSITION

by

Fritz Winckel

Translated from the German by

THOMAS BINKLEY

DOVER PUBLICATIONS, INC.

NEW YORK

Published in Canada by General Publishing Company, Ltd., 30 Lesmill Road, Don Mills, Toronto, Ontario.

Published in the United Kingdom by Constable and Company, Ltd., 10 Orange Street, London WC 2.

This Dover edition, first published in 1967, is a new English translation of *Phänomene des musikalischen Hörens,* copyright © 1960 by Max Hesses Verlag. The author has revised the text and written a new Preface for this English edition, which is published by special agreement with Max Hesses Verlag, Joachim-Friedrich-Strasse 38, Berlin-Halensee.

Standard Book Number: 486-21764-7
Library of Congress Catalog Card Number: 66-28273

Manufactured in the United States of America
Dover Publications, Inc.
180 Varick Street
New York, N. Y. 10014

PREFACE TO THE DOVER EDITION

The positive reaction of American reviewers to the 1960 German edition of this book persuaded the publisher to undertake an English translation. I have revised and updated certain material and it was my intention then, as it is now, to go beyond the area of physical acoustics in music, which is indeed scarcely relevant to the evaluation of the art of music, and to open the field of psychoacoustics from which results the subjective character of musical hearing.

The progress of the last years in the research of psychoacoustics in music has been considered in many aspects: the evaluation of loudness and the dissolution power of the ear; the influence of the acoustical properties of the concert hall on the hearing process; the function of time variation and rhythm in musical perception; the evaluation of the sound spectrum including the unharmonic components. In summary, the improvement of measurement methods has enabled a better description of the hearing process in music.

The analysis of musical structures has often been compared to the structure of speech because the psychoacoustical transformation is the same for both and, in this light, the linguist will find these studies interesting.

The literature in English and German has received equal attention, and it is hoped that composers and musicians will gain a better understanding of the sounding behavior of musical structures by studying the problems treated here.

FRITZ WINCKEL
Berlin, 1967

FOREWORD

In this book an attempt will be made to find the relationships in the laws of nature which are responsible for musical perception.

The more deeply one penetrates into the world of tones, the more clearly one sees that one is dealing with an elastic, finely structured organism, possessing many possibilities of effect. Surprisingly enough, an analysis penetrating to the ultimate atomic particles of the sounding tissue escapes human observation, just as does a submicroscopic analysis in atomic physics and biology. However, there are clearly marked boundaries where sense perception stops, and where recognition by means of mathematical processing of the ultimate vibratory events begins.

The significant thing about studies like the present one is that they consider the whole natural world of sound consisting exclusively in fluctuations of simple formations. This approach is in essence closely related to the peculiarities of the perception process in all living things, and especially to the composition of the nervous system. Compared with such an outlook, the concept of a static skeleton of individual tones, as seen in musical notation, presents only a rough approximation. Further, this approach yields the limits of what still can be called "music."

The author attempted in 1952 to present the fluctuation phenomena of musical texture in their different coordinates in his book *Klangwelt unter der Lupe* (The World of Sound under the Magnifying Glass), and hoped therewith to provide a point of departure for a new music aesthetic. This is a complete revision of that book, in which the scope is expanded considerably. The hearing process is here considered as a perception through the senses which is subject to a complicated psychoacoustical transformation occurring before the real psychological area of perception is reached.

In recent years much new knowledge has been brought to the field of neurology. When we apply this new knowledge to the links in the chain of hearing from the outer ear to the cerebral cortex, we find reciprocal relationships between place and time which yield information in the field of psychoacoustics.

The ideas contained in this book have been influenced considerably by Shannon's information theory of 1948, without, however, dealing with this specifically. Shannon's theory is particularly fruitful for the study of all sorts of communication, an especially interesting branch of which is music. The electrical system theory of Küpfmüller has also found application here. Often it was possible to employ electronic music as experimental proof of acoustical perception phenomena. The electroacoustical musical reproduction in radio and recordings is an excellent object of study, for it has departed considerably from its original goals of ideal sound purity, exact intonation and elimination of the reverberation characteristics of the room, since precisely these factors are undesirable from the standpoint of the listener. It was impossible in this connection to avoid discussing questions of music theory, such as consonance and dissonance, for example.

From such realizations we find hints that can be applied to the performance of music, which not only concern the performer and the instrument but also the listener, and, as a liaison, the room. From this we derive the particular requirements for a concert hall. It became necessary in this connection to discuss the material acoustically, physiologically, psychologically and, last but not least, musicologically.

It is hoped that the composer will find a stimulation here to discover elementary effects in sound emissions and in sound combinations, and that the music listener will be stimulated to penetrate more deeply into the layers of musical perception. An adequate collection of graphs and data will provide the sound engineer and the music producer with support and explanations of the omnipresent phenomena of the effects of sound in connection with performance.

FRITZ WINCKEL
Berlin, 1959

CONTENTS

Chapter I

INTRODUCTION

In the fields of music theory and musicology both simple and compound sound are treated as concrete building material having differing valences. In this way a "function" is ascribed to each building block with respect to the others, whereby a distinctive architecture is formed, possessing a singular tonal character. *Music*, however, is a multiform complex function of sound series, only certain aspects of which have been known to us up to this time.

We have become far too accustomed to compute isolated partial functions schematically and to assume a corresponding validity for the whole, without continuous attentive *listening*, which would test the results. As we shall see, it is not admissible to consider simply the printed value of a note (for example, one having a duration of one second) as sufficient representation of the functional event of this as opposed to other musical notes. The intoned individual sound alone has such a dynamic life that after the lapse of one second both form and necessarily also timbre will have changed. The effect of one sound complex on the following one does not depend upon the intervallic tension alone, but also upon the changes occurring in time during the intoning of the individual sounds. The series of phonemes in language operates similarly.

In music one must take a developmental step now, just as in the field of biology. The consideration of the cell under the microscope as a lifeless isolated component no longer tells us anything of significance. The developing, breathing life of a cell at every instant and the effective power contained in its function within the whole organism are what claim our attention. Not the thing, but the happening is important.

Music is life, it is a living happening. "When does one say that a piece of material lives? When it continually does something, moves...," is the formulation of the atomic physicist Erwin Schrödinger (87). Impulses to movement are, for example, electrical or chemical potential differences. When they are equalized, the tendency to form a chemical bond ceases; temperatures become equalized through heat transfer. Thermodynamic equilibrium results in a condition of constant rest (maximum entropy), a condition which is precisely: death. From the physical standpoint, disorder is continuously created out of a condition of order. Nature strives for the condition of ideal disorder; this was recognized in the last century from the behavior of gases. Schrödinger continues: "The trick by which an organism can keep its place on a rather high level of order consists in reality of a continuous 'absorption' of order out of the surrounding world." We learn further that there are a very few "governing atoms" in germ cells which cause changes in the inherited characteristics of the organism. These are the motif-formers, which give the occurrence its distinctive character.

With such an outlook, we can guess why the effects of music on our feelings have until now remained incomprehensible, and why an analysis of the smallest building blocks fades away before our inquiring senses. We can observe on the larger scale how order is achieved out of disorder, with the end indicated in a perfect final cadence. But the laws of disorder do not obey the causal relationships of classical physics; only from a statistical viewpoint does the tendency of the occurrence reveal itself. We have arrived in the analysis at the border of "indeterminacy," which has become a very exact concept in physics, being more than mere uncertainty or inaccessibility to the human sense organs. This will be dealt with in detail for music.

The building-up of the complex sound, its "becoming," is of significance aesthetically; the end product of the completely built-up sound is less interesting. In listening to the sounds of bells we find much more than simple struck chord relationships. First of all there is the onset of each individual sound; then there is the interference effect of the sounds acting on each other, bringing in new sound qualities; and finally there is the decay of the sound, a breaking-up of the glistening color into the most simple components. This process occurs with the generation of every musical sound; it leads us to observe that with the intoning of written note values additional unintended sound components are generated which can even be unharmonic with respect to the original sound components. Further investigation shows that absolute purity of intonation is very seldom achieved in practical music, and need not even be striven for in order to give the music an experiential content. It is the slight deviation from the "eternally pure harmony" which constitutes the seasoning in the dish, and it should not be set aside.

Therefore, it is not the rigid note value of an unchangeable sound but a richly moving colorful life which generates atmosphere and awakens associations with earlier experiences, which can arouse religious feeling (bell and gong) and stimulate the imagination and the spirit. Especially the latter requirement must be fulfilled by the composer.

"The origin of music, not only in the unique historical sense, but at the same time in the sense of an unending repetition in the realm of every newly awakening individual consciousness, rests on the sound experience. This is the alpha and omega to which music can be traced back" (Arnold Schering). In support of such a statement, not only the musicologist but also the practitioner should have a word. Thus, Franz Schreker, in whose opera *Der ferne Klang* (The Distant Sound) the sound experience is central, says, "Pure sound without any motivic supplement has been used here with caution; it is one of the most important musicodramatic means of expression, a source of atmosphere without equal, which is being used more and more by poets also (among others, Gerhart Hauptmann and Paul Claudel) at decisive moments of the drama. The only thing perhaps more effective is—silence...." "Pure sound," of course, does not mean here a stationary segment of rigid sound, for in another place the composer says, "I disavow the sound which is too clearly differentiable." The tremendous clarity of "definition" of music which is sought after by acoustical engineers in the construction of new concert halls is not in the best interests of the musician.

A sound experience, to be sure, is still not a musical experience. As pointed out above, each building block, as a sound carrier, has a functional value with respect to the other building blocks which are arranged in time before, after and above it. Adorno says in general about this that the material is changed through composition; from this he derives the notion of a materialistic theory of form. "The secret of composition lies in the power which transforms the material in the process of advancing equality." The divergence of sound and music is formulated by Adorno as follows: "Sound, through its being alive, wins a culinary quality, which cannot be reconciled with the principle of construction." He says further in another place, "Each tone that falls into this musical field is immediately more than simply a tone, even though there are no qualities discoverable which are more than those simply of tone" (1).

A structural analysis of music can be approached from different sides. There is the classical description of function based on harmonic and contrapuntal theory; there is also the aesthetic principle of value in the sense of a complete art work. Here the effects of musical happenings are treated quantitatively as information, whereby the statistical method permits recognition of the pregnancy of style characteristics.

With these observations one comes closer to the scientific aspects that expose certain phenomena without which any intellectual analysis of a musical work of art would be incomplete. Especially in investigations in musical psychology there have again and again been mistakes caused by an inexact knowledge of the generative mechanism of sound structures, on the one hand, and of the perception mechanism, on the other.

It is intended here to impart knowledge of this, and thus to explain the physical, physiological and psychological portions of the musical happening. This work should clear a new path for the real aesthetics of musical works of art.

There is general opposition to such scientific methods of observation; however, the reason for this is that such methods have not yet been systematically mastered; people are thus led into the veiled area of metaphysics or into arbitrary emotional interpretation. Although sounds and even more general noise emissions are not visible and not tangible, they are nevertheless physical realities inasmuch as they exist as pressure differences in the air, mechanical vibrations in the middle ear, liquid vibrations in the inner ear and finally as electrical impulses in the nerves leading to the brain. Just as radio waves, light waves and the electrons circling the atomic nucleus are characterized by time and space dimensions, so it is also with sound and other noise emission in all forms.

Therefore, in what follows we will employ a representation of sounds in coordinates of time and place, and we will be surprised to observe on the macroscopic border, as it were, that an absolutely precise location in place *and* time, that is, frequency (pitch) and time, is not possible, a situation which obtains also in atomic physics. From this we can derive some information about the "purity" of sound, the exactitude of intonation and further phenomena. The descriptions of sound phenomena here, in accordance with the above considerations, make use of a time continuum (Chapter II) or a coordination of the partial components (frequency scale as in the spectral representation in Chapter III)—as the actual local coordination in the inner ear and nervous system. In reality an influence is exerted on the frequency structure of the sound by the duration limitations of sound—sound onset, sound change (modulation) and sound stop; thus, the sound quality as a certain relationship of partial vibrations will be changed. Paul Hindemith correctly points out (34) that one must not think of music as a series of single emitted tones, but as a continuum. In the flow of musical events energies are released which penetrate into the perception center of the listener by means of the psychoacoustic transformation of the hearing mechanism; of course, this has an analogy in visual perception. The natural laws are just as valid for such effects of energy distribution as for the processes of energy in atoms, in the cellular aggregate of living beings, in the flowing of water and in the movements of electrons.

We shall by no means favor a materialistic philosophy in our approach, for the processes released through music in the area of the soul will remain untouched. The goal of these investigations can be thought of as attained if certain phenomena in the reproduction of music and its effects on hearing are explained in an exact way.

Chapter II

STATIONARY SOUND

1. Basic Concepts

We see from the introductory chapter that the following situation obtains: a single intoned sound is meaningless in the sense of a musical investigation when, after onset or any other change it becomes stationary, or reaches a value which remains unchanged in time. If we investigate this stationary acoustical phenomenon anyway, it is only with the intention of establishing a basis which later on will make the remarkable psychoacoustical laws of non-stationary sound more easily understood. Besides, we may state in advance that in the laws of nature there is a relationship between the stationary sound structure and that of sound movement, or the onset process.

In musical acoustics the term "tone" [German: *Klang*] has until now been defined as a sound emission whose character is dependent upon the pitch, loudness, duration and compounding of fundamental tones and overtones, whereby the number of vibrations per second of the overtones is a whole multiple of the fundamental. The whole-number overtones are referred to as partials or harmonics of the fundamental. In an increasing series (first, second, third,... partial) the intensity of the partials progressively declines. The sound character depends, in addition, on the onset process, which will be discussed in Section 2.

2. The Partial Spectrum

Since the introduction of acoustical spectral analysis (about 1928) the partial components of a sound have been diagrammed symboli-

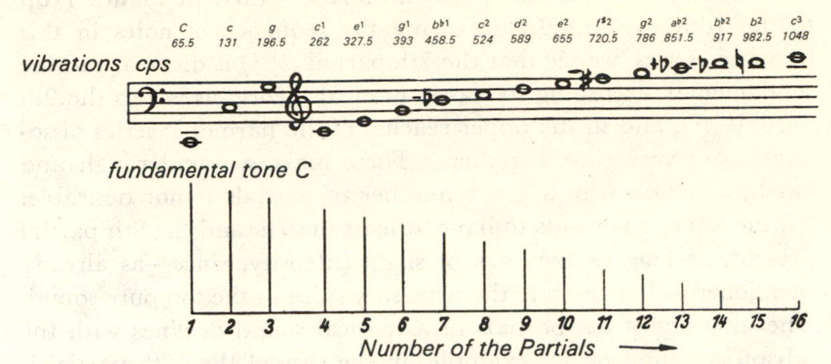

FIG. 1. Harmonic spectrum of a sound up to the 16th partial.

Above: partials in musical notation (harmonic series) with vibrations per second; below: theoretical partial spectrum. The 7th, 11th and 14th partials are a little higher, the 13th a little lower than shown in the notation.

cally in the form of a sound spectrum, in which every line represents a partial (sine wave) and the length of the lines its intensity (Fig. 1).* Between these lines, which as symbols of the harmonics are equidistant from each other, there are no other sound components. As opposed to this, "*noise*" gives a continual spectrum in which all possible partial components, also the inharmonic ones, follow closely after one another (Fig. 2). Noise can be grumbling, metallic, hissing, etc., for which the acoustical energy emphasizes accordingly the low, the high or the very highest frequencies. Because of the formation of such sound colorations one speaks of "colored noise," to which we will return later. If, however, the sound energy is distributed equally over the entire audible range of the frequency band, one speaks of "white noise"—again an optical analogy—for the impression of white light originates through the combination of all colors of the optical spectrum.

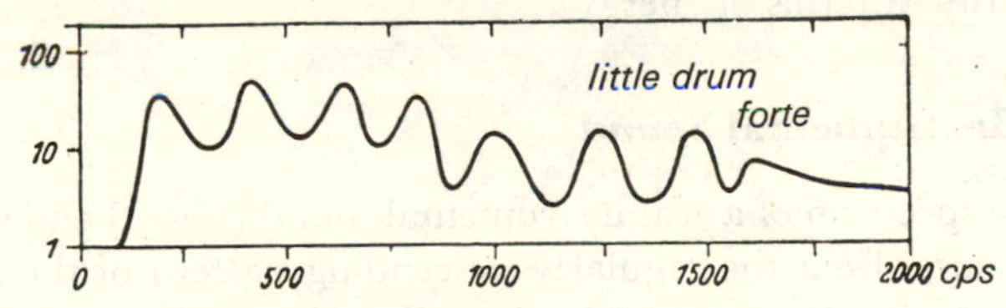

FIG. 2. Noise spectrum of the drum.

The space below the envelope should be thought of as crowded with spectral lines.

* The German system of musical notation is used throughout this book.

The theoretical partial spectrum is represented in Figure 1 up to the 16th partial. In observing the sequence of notes in this harmonic series, we see that the 7th partial ($b^{\flat 1}$) is dissonant to the fundamental, according to classic musical theory, as is also the 9th partial (d^2), and in the upper reaches of the harmonic series dissonances are very close together. These make a sound rough and harsh, a reason why a great number of partials is not desirable. All the same, one wants to have at least the 7th and the 9th partial present, as long as they are of slight intensity, since—as already mentioned—they provide the necessary spice in the too pure sound. The intensity of the partials in a musical sound declines with the advancing number; for example, in the case of the 16th partial it declines to about one per cent, which is a decrease of about 40 decibels. With sounds in the middle range (70 phons) with a fundamental above *e* only about the first 10 or 12 partials contribute to the formation of the sound. The author has observed this in the case of some of the greatest voices (Caruso, Gigli, etc.) as well as with instrumental sounds.

Classical music theory runs into difficulty in attempting to derive the scale from the harmonic series, because of the insuperable problem of trying to accommodate the highest partials in the diatonic scale. It may seem peculiar that the harmonic series seems harmonic in a physical analysis (compare the numbers of the frequencies), that is, consists of whole multiples of the vibration of the fundamental, and yet a diatonic analysis appears sometimes consonant and sometimes dissonant. The explanation of this phenomenon to the effect that the ear perceives the harmonic series in a logarithmic condensation, as for example the octave-shrinking in piano tuning, is not sufficient. From the spectrum in Figure 1 one can see how slight is the significance of the partials in the advancing series in terms of energy.

3. Instrumental Sound

The spectrum of a real instrumental sound (Figs. 3 and 4) differs, to be sure, from the regularly descending pattern of the spectrum in Figure 1. One notices in the series of partials that some are of greater intensity than others. Their function in coloring the sound will be discussed below.

In the schematic application of these concepts there has been a

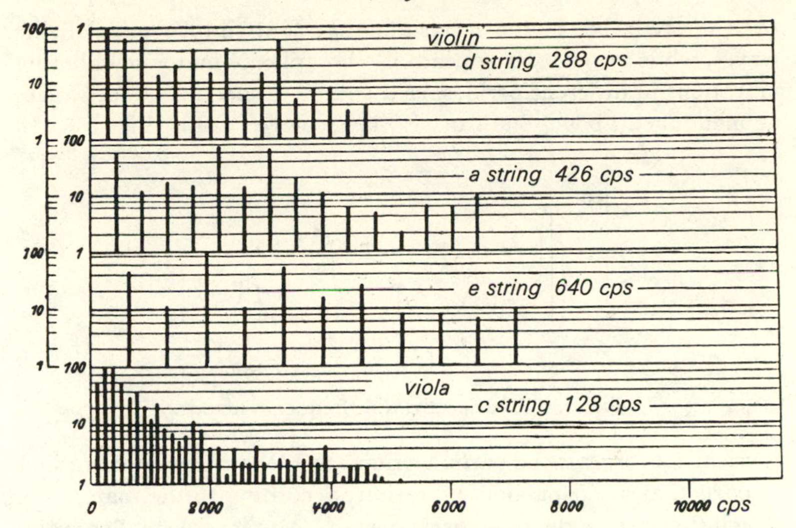

FIG. 3. Spectra of partials of violin (above) and viola tones (below) (according to Meyer and Buchmann).

growing tendency to ignore the physical manner in which a sound originates, and even in the study of harmony a one-sided viewpoint has finally been established. A single partial is symbolized simply

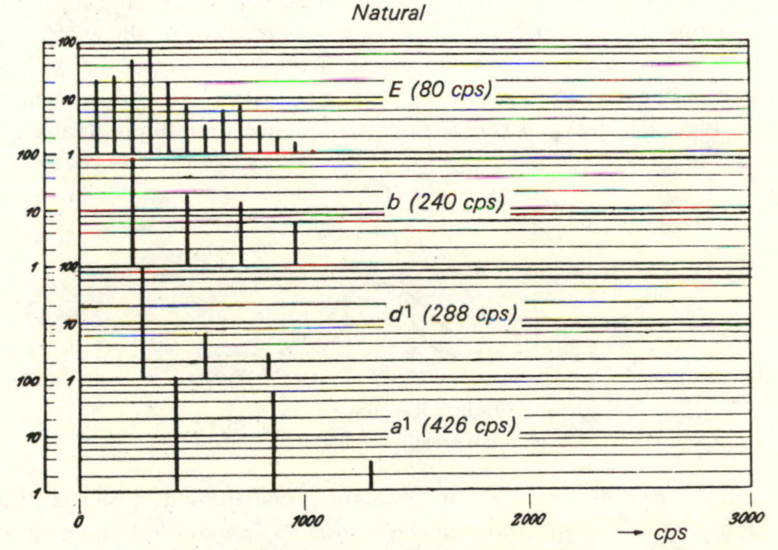

FIG. 4. Spectrum of natural horn partials.

by a discrete line in the spectrum, that is, a simple series of sine waves, while in reality it appears as a complex of many neighboring vibrations in different periods, as a complete spectral band, which, as mentioned above, one can also call colored noise (Fig. 5). In

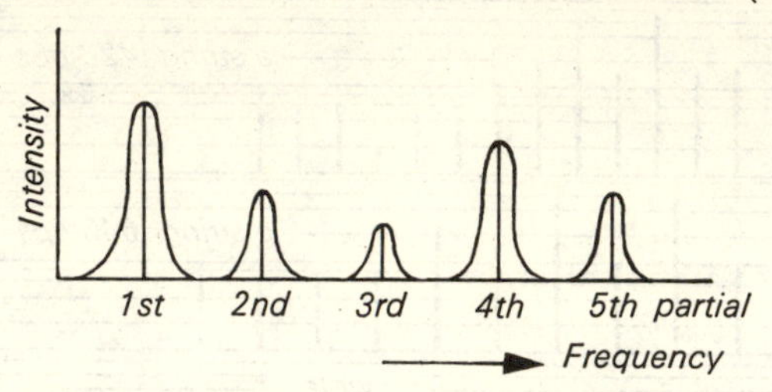

FIG. 5. Real spectrum of frequency bands.

musical instruments the partials originate as the resonant vibrations of parts that are capable of vibration; according to the material of which they are made, these parts permit a larger or a smaller group of vibrations per partial, according also to the degree of damping in the resonant system. The envelope of the resonant vibration in the area of one partial is called the "resonance curve." The line spectrum is thus only a simplified representation of the real behavior of vibrations. The steeper the resonance curve, that is, the sharper the resonance, the less damping there is. This characteristic depends in turn on the molecular characteristics of the vibrating materials (mass and elasticity), and in the case of electrical vibrations it depends on the choice and dimensions of the circuit elements. In Figure 6 curve *A* shows sharp resonance and slight damping,

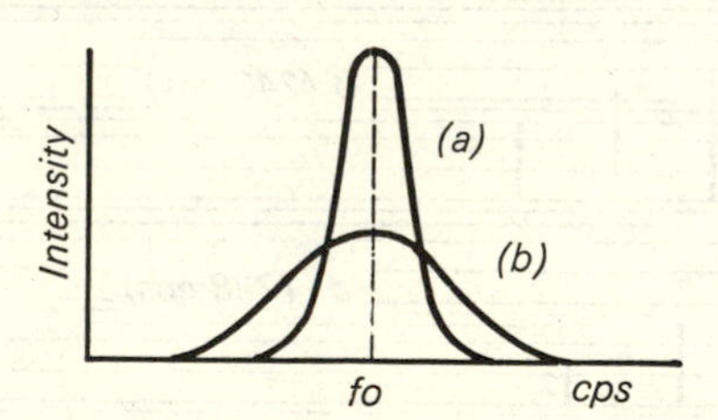

FIG. 6. Resonance curves.
(*a*) : little damping (*b*): great damping; f_0: resonant frequency.

while curve *B* shows slight resonance sharpness and substantial damping. The spectral line, belonging to one stable frequency of infinite duration, represents the exclusively theoretical case of zero degree damping; thus, a single discrete spectral line arises without further neighboring lines. A musical tone can thus be thought of as a "colored noise," a term to be dealt with in detail below. For perception via the nervous system other laws are valid, as will be discussed in Chapter VI.

Resonant vibrations can arise in a hollow body if its walls are excited or if a vibrating air column of the resonant frequency of the hollow body is permitted to enter through an opening. If the inner surface is hard as metal, the resonance is sharp; if it is covered with an absorbent coating, the resonance is less sharp and the resonance curve is broader. The damping of the soft inner surface of the human mouth causes larger losses through absorption than the hard inner surface of a violin body.

Such hollow bodies are called Helmholtz resonators. The frequency range in which overtones from the sound source receive a reaction depends on the size and to some extent on the shape of the hollow body. The sound of the human voice, for example, yields a spectrum whose components result from a number of cavities: pharynx, oral cavity, nasal cavity and windpipe. The mucous membrane covering the nasal cavities yields a greater damping effect than, for example, the area of the hard palate.

The theoretical representation of the spectrum as a series of "harmonics" must permit a further modification, for an exact spectral analysis of instrumental sounds shows that significant frequency variations of the whole-number relationships of the overtones occur with all instruments, so that one can scarcely speak of "harmonics."

A drastic case of this has been produced in an experiment with organ pipes by J. Meyer (63). The resonant places of the pipes were excited by a gliding sine wave from a loudspeaker. The frequency deviation from the whole-number multiples of the fundamental is represented in Figure 7 for three different diapasons (*Mensuren*): *Zartgeige*, *Principal* and the distant *Nachthorn*. With the 10th partial there is already a 50 per cent discrepancy between the resonance frequency and the fundamental frequency; with the *Principal* there is a 100 per cent discrepancy, so that the resonance for the 10th partial excites the 11th partial; for the *Nachthorn* the 100 per cent discrepancy is arrived at with the 4th partial. Similar discrepancies can be discovered on most wind instruments, for which R. W. Young has provided measurements (116). Even for piano strings there are systematic frequency discrepancies of the harmonics, and as a matter of fact, rather uniform discrepancies, regardless of manufacturer, at $c^1 = 256$ cps; the discrepancy for the 2nd partial is 1.2 cents, with a regular doubling after the 8th partial—agreeing,

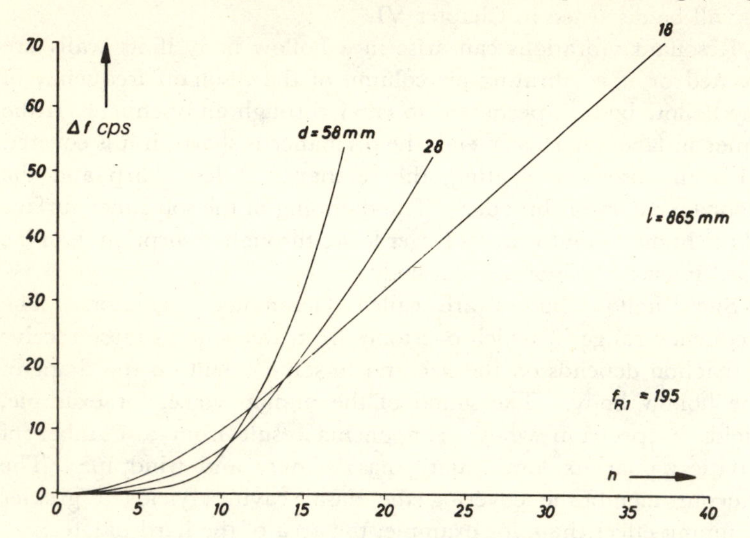

FIG. 7. Frequency deviations Δf of the resonances of organ pipes from the harmonic partials of the fundamental resonance f_{R1}. n = number of the partials; d = diameter; l = length of the pipes (according to J. Meyer).

by the way, with the theory of stretched strings. The proportionally increasing discrepancy indicates a spreading of the line interval in the spectrum, according to which the formation of difference tones would indicate a lower fundamental than the 1st partial really is.

4. Formant Theory

Pronounced resonance peaks in the partial spectrum contribute greatly to the recognition of the tone character and to the establishment of the tone color. Such pronounced resonance peaks were given the name of formants by L. Hermann (1890); they are symbolized by means of spectral lines of relatively greater length. For example, the vowel sound "o" is characterized by its preference for resonance in the frequency area of 400–600 cps, the vowel sound "ă" in the area of 800–1200 cps (cf. the formant table, Fig. 8 and the musical notation of vowel formants, Fig. 9*a*). Basically, every vowel is formed out of two formants, to which still a third formant and even some other formants of higher order are added; but because of their slight significance, they will not be discussed here. The relationship of the upper and lower formants can be seen in Figure 13. In Figure 8 and 9*a*, for the dark vowel sounds (u, o, å) only one formant is given, simply because the second formants are of very slight intensity. In the case of the vowel "a" the two formants lie so close together that they cannot be separated according to frequency.

With this viewpoint the disagreement between H. Helmholtz and L. Hermann (1890) can be set aside. The question was: Are the

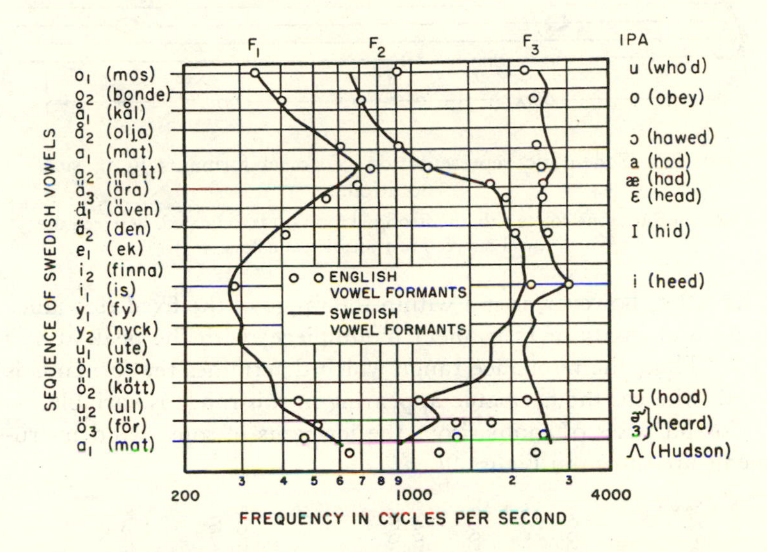

FIG. 8. Formants F_1, F_2, F_3 of Swedish and American English vowels.

See Fig. 73 for an explanation of the phonetic symbols (according to G. Fant).

formants built up harmonically to the fundamental, or are they unmovable and absolute frequencies which can also be inharmonic to the fundamental? Both concepts are valid if one assumes that a formant is not schematically a discrete frequency but a frequency *range*. If, for example, a sound with the fundamental of 100 cps is intoned, a frequency range with a resonance peak in the third harmonic, 400 cps, will vibrate in resonance (Fig. 10). If the pitch is then raised to 105 cps, a harmonic of 420 cps will be excited.

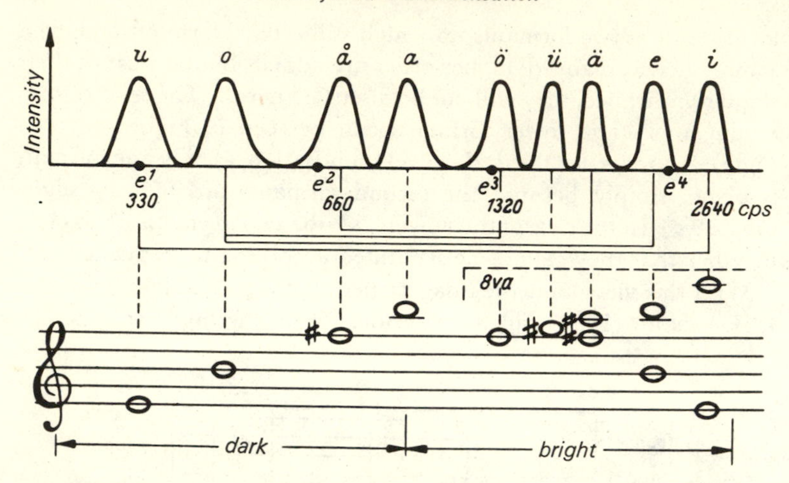

FIG. 9*a*. Schematic representation of vowel formants in musical notation.

For the dark tone colors only the strongest formant is indicated. Between the vowel curves are transitional colors.

The latter, however, is still within the range of the fixed resonance (e.g., oral cavity or instrument mouthpiece) with the peak at 400 cps. Thus, the resonance range which determines the formants is fixed, whereas the harmonic appearing in this range is variable.

For purposes of comparison, the formants of some wind instruments are shown in Figure 9*b*.

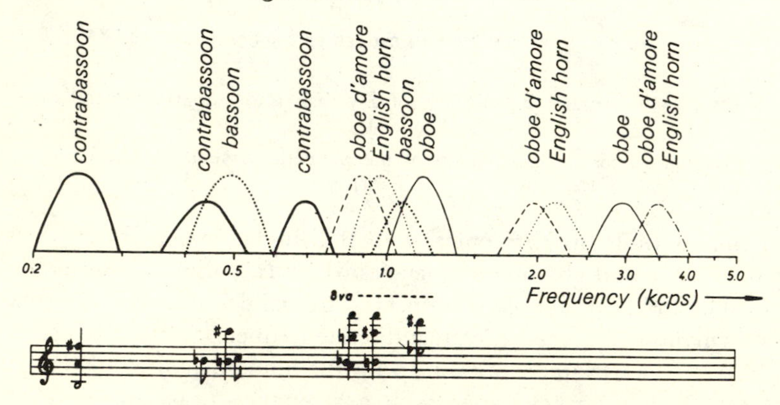

FIG. 9*b*. Formants of some wind instruments (according to J. Meyer).

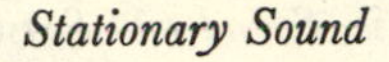

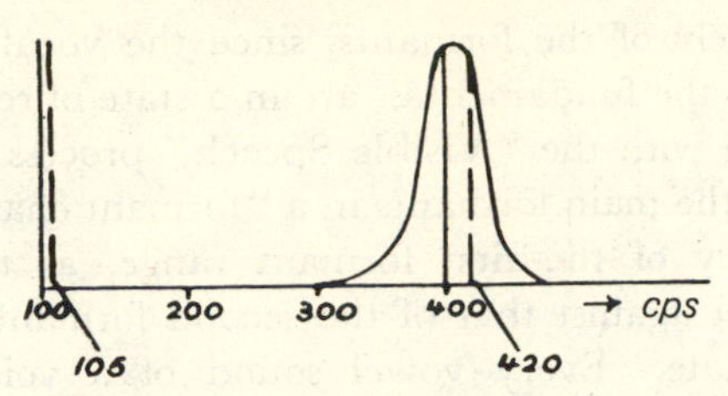

FIG. 10. Influence of formant area on partials: the 3rd harmonics for the fundamentals 100 and 105 cps.

One must be cautious about employing discrete spectral lines schematically to represent overtones and especially formants. One comes closer to the physical situation when one presents the limiting envelope of the spectral lines, which lie close together (Fig. 11).

In what follows we will limit ourselves to a discussion of the formants as the most significant earmarks of sound. In the investigation of spoken as well as of musical sounds it has been found that in the recognition of tone color there are chiefly two humps in the integrating envelope which are of importance; these correspond to the formants (Fig. 11).* Naturally, for finer sound differentiation

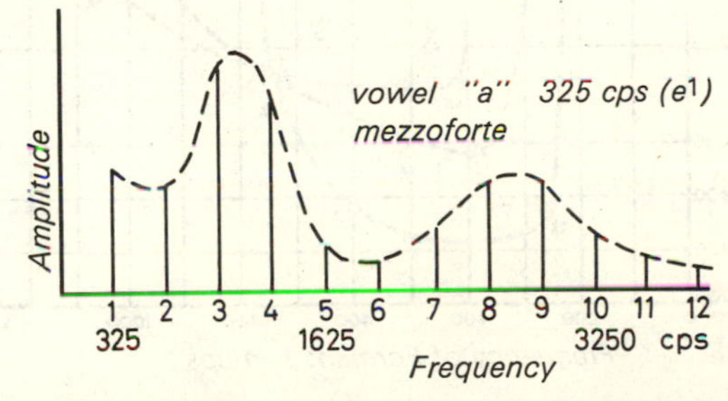

FIG. 11. Envelope of the spectrum with upper and lower formants.

many other factors are important, e.g., the onset transients, time change of formant frequency, loudness, pitch and so forth. These influences we will ignore for the present in order better to see the function of the two main formants. It is possible that the human brain, in order to differentiate a sound, scans the envelope of the spectrum as if following along the outline with a pencil (Potter) (77). A point of reference for this is the case of whispering, which

* They are not always clear, as in the case of the natural horn (Fig. 4).

consists exclusively of the formants, since the vocal cords, which normally supply the fundamental, are in a state of rest.

While dealing with the "Visible Speech" process (79) someone entered both of the main formants in a "formant chart," where the middle frequency of the first formant range, as the horizontal coordinate, is set against that of the second formant range as the vertical coordinate. Every vowel sound of a voice analysis is marked with a dot. In Figure 12 the man's vowel "I," for example,

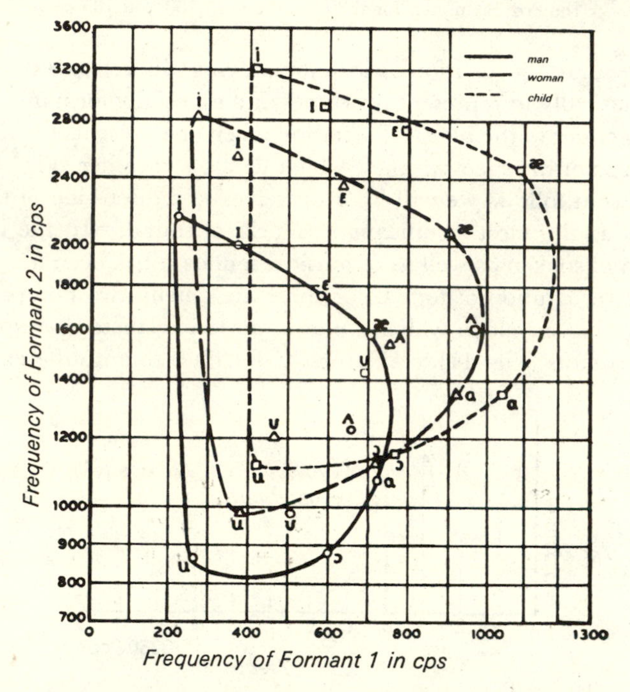

FIG. 12. Vowel triangle or formant chart of man, woman and child with formants 1 and 2 (according to Peterson and Barnay).

is marked with a dot at the two formants, 380 and 2000 cps. When all of the vowels are entered on a formant chart, as in Figure 12, the complete range of vocal color will approach the shape of a triangle. Surprisingly, this representation is similar to the vocal triangle established by Hellwag in 1781. The time sequence of diphthongs is represented in Figure 13, that of speech in Figure 14*a*; every 20 milliseconds a point is entered representing the formant value at that moment. Every dot on the chart is characterized by a specific tone color. For use in practical music the grid would have to be logarithmic, analogous to a musical scale. Thus we obtain a *meter*

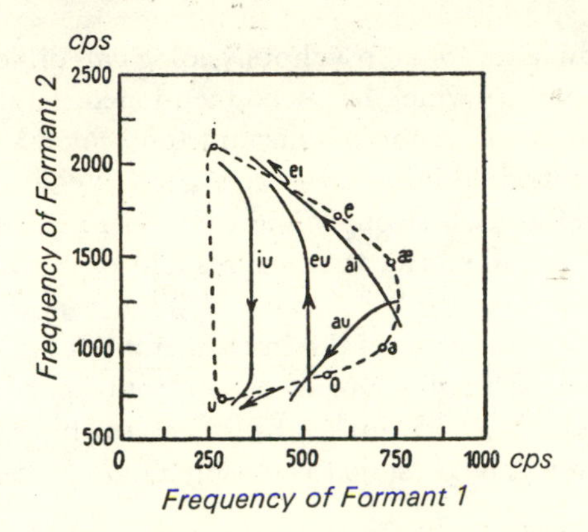

FIG. 13. Diphthongs in the formant chart (according to Potter and Peterson).

of tone color which, through division of the chart into diamonds, registers the size of the still barely perceivable step-changes in tone color. The threshold of perceptibility for vocal tone color lies around a semitone.

This was established statistically by asking a number of people when they recognized the point at which one tone color faded into a neighboring tone color. This gives us a metrical field with a certain number of sensitive locations as functions of complex

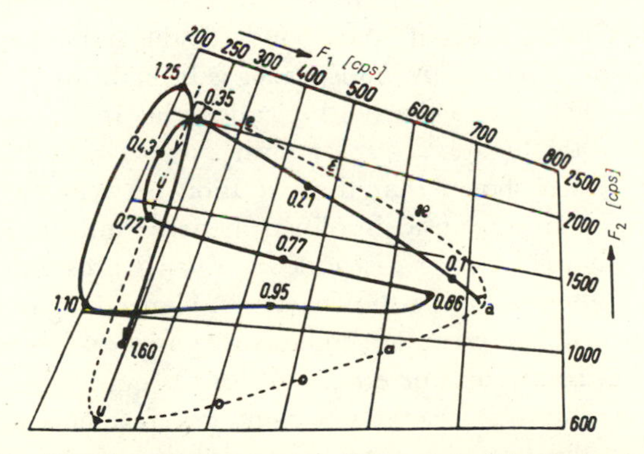

FIG. 14*a*. Tracing of the sentence "I love you" in the formant chart (according to G. Fant).

acoustical excitations. These valences—which one can also apply in sound psychology, according to C. Stumpf (99)—do not lie close together; tone colors do not form a continuum in our perception, as one would tend to assume.

There is an additional psychophysiological difficulty: this cell structure of valences which has been mentioned is only approximate for a stationary comparison of constant tone colors. The course of the tone color modulation, that is, the speed of change of the stimulation, determines how strongly the valences in a metrical field are distorted, or how the grid at the threshold of perception becomes finer or coarser.

Although the tone color of the instruments of our classical orchestra, including the singing voice, occupy only part of the cells of our color field, electronic musical instruments have filled in many gaps, so that it will be possible to develop a more systematic ordering of tone color.

The results of a tone color analysis can be tested with the help of a "sound synthesizer," developed first in the United States and later improved by G. Fant in Sweden (see Fig. 14*b*). Using a pointer, one scans, "approaching" the desired tone color. The pointer controls corresponding color divisions in the vowel triangle through manually controlled potentiometers, which then undertake the filter adjustment for the desired formant frequencies. A sawtooth generator is connected to the circuit, which in imitation of the sound production in the larynx generates a sawtooth voltage which passes through the above-described formant filter circuit, producing the vowel spectrum with the desired envelope (e.g., Fig. 11). The filter chain corresponds to the human vocal tract (organ of articulation). A loudspeaker is brought in at the end of the circuit which makes the synthetically generated vowel audible (analogous to the function of the mouth).

Experience has shown that the variation between two formant frequencies F_1 and F_2 in a simplified model is sufficient if a third fixed formant frequency F_3 is added. Even words can be made distinguishable by scanning the chart as in Figure 14*a*, since through the movements or interruption some consonants can be simulated.

Such a device can also be employed for the generation of sounds for electronic music. A composer could generate new sound series by scanning the formant chart, altering pitch by means of the generator.

Basically this circuit is what F. Trautwein in 1930 realized in his instrument the "Trautonium" (101), in which the player glides his finger along a resistance wire on a manual to vary the pitch. The inventor and also Oskar Sala have made extended studies of vowel color and general tone color formation. In 1948 Sala described how important a second formant is for the stabilization of the tone color (84). He demonstrated with the last bars of the Brahms violin sonata in G that it is not possible with only a simple electric overtone circuit having a single variable formant, to bring about a

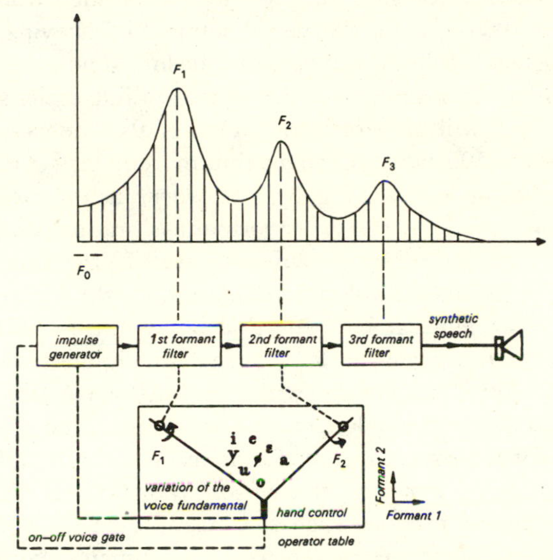

FIG. 14*b*. Manual operation of the Swedish speech synthesizer of G. Fant (according to G. Fant).

change of expression within the bounds of a given basic color, from a muddy minor into a bright light. Every change of the resonance circuit which causes a change in the formants has the effect of a basic change in color, for example sliding u–o–a–e without finer variations. Only with the addition of a further resonance circuit creating a second formant did the sound structure acquire sufficient stability; the adjustment of the amplitude of only one of the formants produced the brightening or darkening of the tone color.

After one has grasped the physical principles of the formation of tone color, it is a simple matter to vary the nuances of expression, which have been few and accidental up to now, and to extend the area from dark to bright. This observation is not limited to completely new electroacoustically generated tone colors, but can be employed in the future, with appropriate circuitry, for the tone colors of the existing instruments.

With this simplified theory one can explain in general why the tone color of musical instruments and the human voice changes with the pitch, the tone color becoming brighter with ascending pitch, or even showing a tendency toward colorlessness. Let us assume that a soprano voice runs through its total range from c^1 to c^3 (about 250–1000 cps) on the vowel sound "o," having the main formants between 400 and 600 cps: On the lowest note the excitation of the formants is possible over the entire audible scale, so that the tone color "o" will be perceived. That is also the case when the pitch lies about 500–600 cps, the maximum frequency of the formant area of the vowel "o" (see Fig. 8). However, as the pitch ascends, the functioning area of the lower formant is gradually left behind, and the "o" color begins to lose clarity. If the soprano voice ascends to c^3, it will be able to stimulate only the formants of the bright vowels. Thus it happens that a coloratura in the upper regions of her range can no longer sing intelligible texts, but can make only flute-like sounds with an "ih" character. Similar observations can be made for musical instruments; here the change in color is not so strong, since the formants occur both higher and in greater numbers. For the latter reason, the clarinet (Fig. 15*a* and 15*b*) and bassoon produce such significant tone color differences over their range that one can speak of instruments having several registers. The unsatisfactory psychological approach to the relationship of pitch and tone color, which has been discussed by Jacques Handschin among others (32), now can be replaced by the above clear approach from the physical standpoint.

The characterization of sound essentially through two formants is of great significance, even though it is not organically adjusted to the hearing process; this is because we do not distinguish single resonance areas of the spectrum, as does the keyboard of the bulging basilar membrane in the inner ear, but rather hear an integrated blending. However, there is no other characterization of tone color known. In addition, it might be pointed out that some phoneticians with a great deal of hearing experience have been able to hear single formants in a sound; and further, it is possible

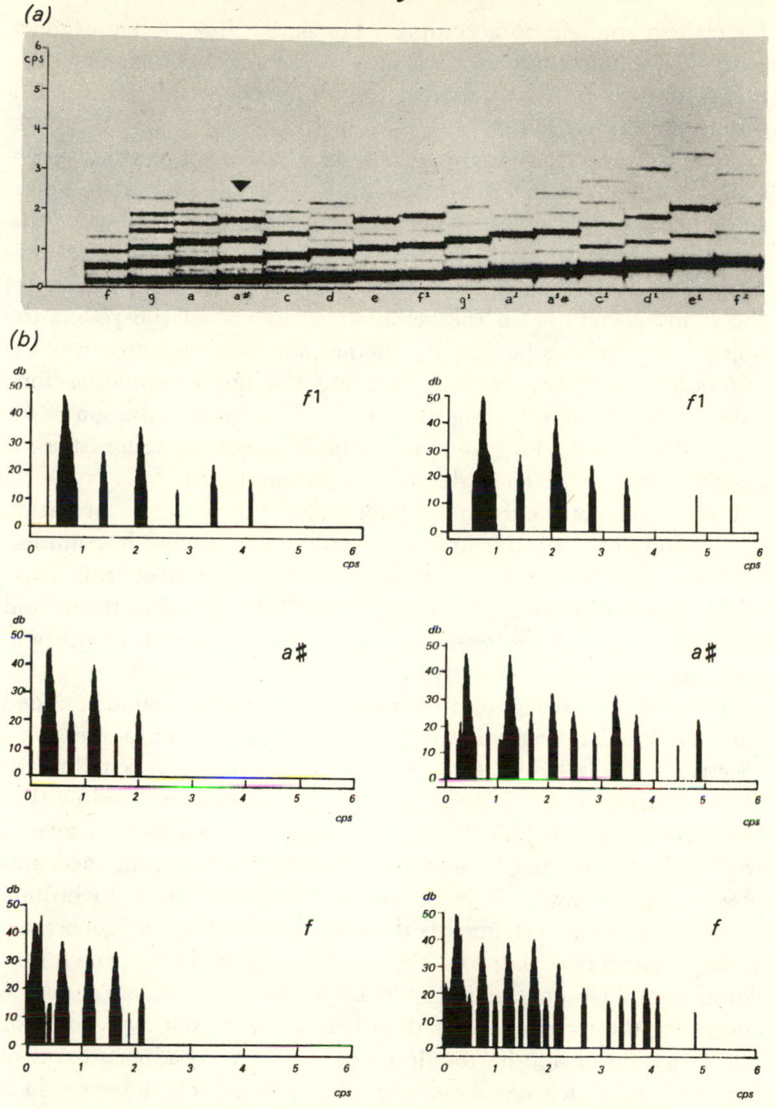

FIG. 15. Spectrogram of sounds of the clarinet produced with the Sonagraph.

(*a*) Scale of clarinet sounds along the abscissa, partials in different layers (thick lines are formants). (*b*) Spectra taken from sounds f, a#, f¹.
Left: legato. Right: staccato.

for trained students to accomplish this under given circumstances, namely, if a tone of the formant pitch is sounded beforehand. The ability to hear in this analytical way requires special talent and a natural predisposition.

A further criterion for the formation of tone color is the type of sound source. In this sense, the vocal cords of the human voice, the bowing of the violin string, etc., must be considered as *impulse generators*. The strength to which the formant frequencies are formed depends upon the relationship of the vibrating frequency of the sound generator to the resonant frequency of the resonating object, and depends further on whether, for example, only the odd-numbered overtones respond, yielding the hollow-sounding tone color. There are still further influences, e.g., the distribution of the main energy in the low, middle and high ranges, characteristics for which have been given in detail by C. Stumpf (99).

On the basis of the thorough knowledge of the spectral formation of the sound of musical instruments (60) as well as vowels, countless attempts have been made to produce these sounds synthetically from sine waves of different frequencies. It was thought that this would be especially easy if the sounds selected were of constant intensity in time, such as those produced by organ pipes.

It was a great disappointment that the synthetic sound sounded very unlike the original, although displaying exactly the same spectrum. This led to the suspicion that there was a failure to observe variations in the intensity of single partials, or possibly that the harmonic spectrum of constant tones is supplemented by several weak inharmonic partial components which have remained unobserved until now. It is interesting that it is easier to imitate the sound of brass instruments than sounds produced by generators made of softer material, e.g., the vowels. In the case of the severe damping of the human articulatory organs there are wider resonance curves for the formants, and therefore a wider excitation zone for the formation of non-harmonic partials. A. W. Ladner (48) is of the opinion that the low-energy inharmonic components lack significance in high-frequency ranges, as the ear is not so sensitive there. Hence one must look for these characteristic components in the area of the lower tones, possibly even in the *subharmonic tones or their overtones*, for the undertone row is not harmonic to the overtone row (Fig. 16).

Oskar Sala has made use of the subharmonics to achieve special

FIG. 16. Overtone and undertone series.
The numbers indicate the intervallic relationships of the tones in the series.

effects in his sound studies for the mixture trautonium (84). Figure 17 shows how one can excite three subharmonic voices from one main generator.

These closing remarks show that the clearly observable spectral

FIG. 17. Subharmonic mixtures.
Example from a composition by Harald Genzmer, *Concerto for Mixture Trautonium and Orchestra*. MG = main generator (played tones); AG = auxiliary generators (accompanying tones).

representation of musical sounds, which has been so firmly introduced, definitely does not describe the character of sound completely. In addition to the above objections, there are further factors to be considered. These, treated below, constitute the major statement of this book.

Chapter III

THE ONSET BEHAVIOR OF SOUND

1. On the Inner Processes of Movement in Sound

Coming now to real and live sound in its entirety, as we employ it for musical assessment, we must draw attention to a peculiar and significant phenomenon of sound, which depends upon the character of the sound generators or the resonant bodies. This is the *onset process* of the sound which, just as in the above case of the resonance curve, is dependent upon the damping of the vibrating body. In order to understand this, we must investigate another condition normally taken for granted in musical acoustics.

According to the mathematical vibration analysis of J. B. Fourier (around 1800), a discrete partial of sound, which by definition must have sine wave characteristics, cannot exist unchanged unless the sound is of infinite duration. The *sine wave* observed in practice must be represented exactly as in Figure 18; that is, out of a condition of rest lasting indefinitely long, it must begin at a point of time t_1 and conclude at a later time t_2. The continuity of the three events, namely the states of rest before t_1 and after t_2, as well as the constant vibrating process from t_1 to t_2, is disturbed by the sudden changes at points t_1 and t_2, and hence an analysis of the sound according to Fourier cannot be absolutely correct. In practice, the "sudden unsteadiness" shown at the instant t_1 and also t_2 is not possible because of the inertia of the body to be set in motion.* However, even an approximately sudden switching of conditions

* One can get an idea of this experimentally by recording a sine tone on tape, cutting out a single oscillation, pasting it onto blank tape and listening to it.

leads to a disturbance of the stationary condition; this disturbance can be observed over many periods, and theoretically throughout the complete vibration process. If the vibration is an acoustical one, the initial disturbance will be perceived as audible distortion. Strictly speaking, there are no exactly periodic events in our life on earth, since every technically generated vibration must have been switched on at a finite point in time, leading to distortion.

The mathematically defined sine wave exists only as a theoretical concept. If there actually were one, having begun before our birth and continuing into infinity after our death, we would not be able to notice it, just as we would not notice the blue of the sky if it were never disturbed by clouds and the darkness of night.

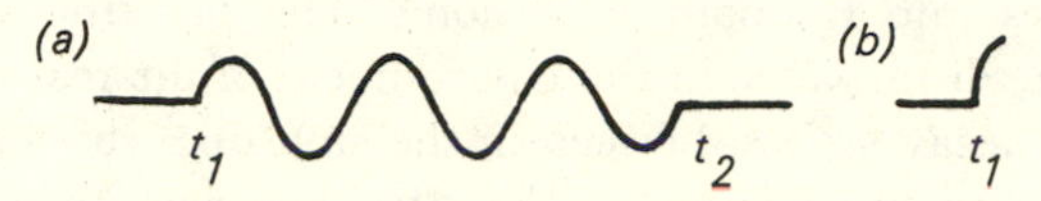

FIG. 18. (*a*) Vibration event limited to the time $t_1 \ldots t_2$. (*b*) "Inconstancy" of the vibration onset at t_1.

The considerations in the first chapter have only an approximate validity. It would be more correct to speak of "quasi-stationary events" and of "near-sine wave" vibrations.

These considerations have basic significance for the further observation of the effect of music, even for the understanding of the life process in general. There is certainly no organic development in our world which has neither beginning nor end! Birth and death are the discontinuous functions that have great weight in the functional course of the life curve between the two end points. The life curve, as opposed to a periodic repeatable curve, is characterized as a singular event by the spectral expansion in the sense of "noise." The individual is not only a structural unit but also one enclosed in time. Beginning and end are part of its being (33). Our historical phases and intermediate stages are also mathematical-physical phenomena which are perfect and isolated in themselves, where there are no constant transition functions. A musical work in the variety of its form can be considered analytically in this way as a perfect and complete whole, and thus becomes an immediate image of life.

But, to continue with our investigation of sound, the more suddenly a system is excited, the more suddenly a sound begins, and the more a disturbance in the existing condition shows itself in the enrichment of the spectrum with a large number of vibrations of all frequencies following closely one upon the other. The limited wave train, as represented in Figure 18, appears as a "group of waves," as it is described physically. For the theoretical case of an absolutely sudden switch from zero to maximum value vibration—perhaps by means of an electrical switch—there occur in the first instant infinitely many vibrations of different frequencies following closely after one another, and this is perceived as sudden noise (click). Let us take as an example the sound from a piano which is suddenly excited, struck at the key a^1 (440 cps), as presented in Figure 19. The noise spectrum of the onset event for the fundamental of the struck string has a concentration in the area of its frequency f_0, as this comes into resonant vibration. The parasitic vibrations caused by the disturbance in the environment of the real frequency f_0 gradually decay, as can be seen in the envelope which gradually becomes smaller and narrower (Fig. 19); however, they arrive at

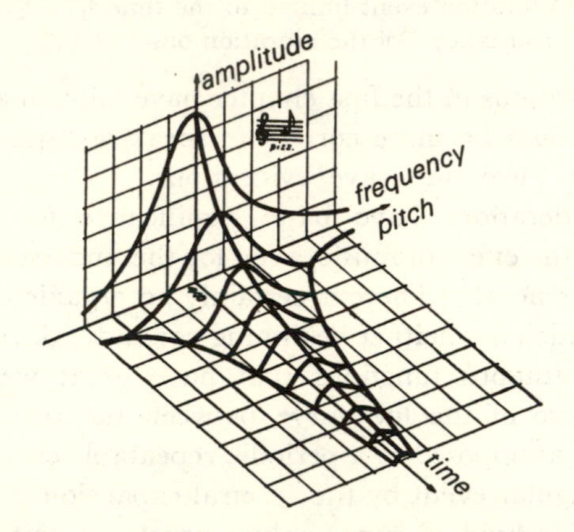

FIG. 19. The decay process of the fundamental a^1 of a struck string in a perspective graph.

a zero value only after a very long time, in the case of an undamped vibrating system only after an infinite amount of time. The noisiness of a sudden onset of a sound can be heard very clearly when one intones a vowel with a hard glottal stop. As opposed to this, a soft, pressureless, swelling vowel sound begins noiselessly. In the case of the hard attack we have a situation involving physical transients, not simply a mechanical noise of the opening and closing of the vocal chords.

Along the time axis of the spectrum in Figure 19 one hears essentially the saddle-form of the figure, that is, the decay of the tone of 440 cps. This does not entirely represent reality. Actually, the sound does not reach maximum amplitude immediately at time t_0. Just before the descent in the saddle-line one must imagine a short ascent on the other side of the frequency coordinate, which has been omitted here to simplify the representation. Besides, the figure is valid only for the first partial, whereas in order to characterize the sound as a whole, one must imagine a similar envelope for each of the remaining partials.

The objectively ascertainable noise caused by the sudden onset of a sound is not so strongly perceived subjectively, for there are protective devices in the ear which tend to dampen such disturbances. Chapter VI deals with this in more detail.

2. The Type of Onset

K. Küpfmüller has presented an exact mathematical proof of the spectral frequency-spread phenomenon at onset of a vibrating system (45). What he has derived especially for electrical vibrating circuits we can employ in the observation of acoustical vibration. In doing so, the action of the hammer of the piano in striking the string, the initial excitation of an organ pipe through the column of air, etc., are to be regarded physically as the onset of the vibrating system, inasmuch as the system, which up to that time has been at rest, has been energized.

For mathematical treatment, let us begin with the concept of a periodic right-angle vibration (Fig. 20), about which we know by

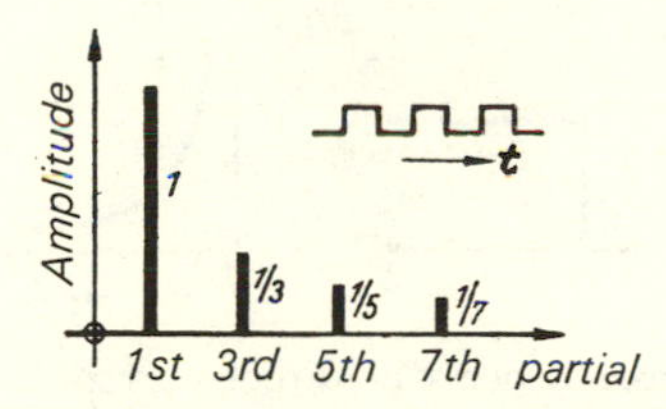

FIG. 20. Periodic square wave with the corresponding spectrum, interrupted after the 7th partial.

Fourier analysis that it can be dissected into an infinite number of sine waves of varying frequency, namely into a root f_1, the 3rd,

5th, 7th, etc. partials, *ad infinitum.* Figure 21 shows that by superimposing the 1st, 3rd and 5th partials with corresponding reduction in amplitude by 1, $\frac{1}{3}$ and $\frac{1}{5}$, a right-angle wave form is

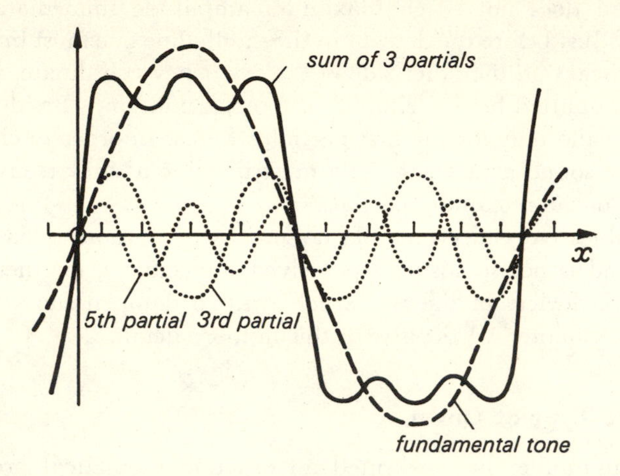

FIG. 21. Approximation of a right-angle wave using the 1st, 3rd and 5th partials.

approximated. The further addition of odd partials f_5, f_7, f_9, etc., *ad infinitum*, would yield an exact right-angle form. If one now imagines that the period T of the right-angle vibration is infinitely large, one arrives at a function which is exactly that of the onset function of direct current (Fig. 22). Therefore, in this case also an infinite number of sine wave partials of the odd numbers are formed, as shown on page 26.

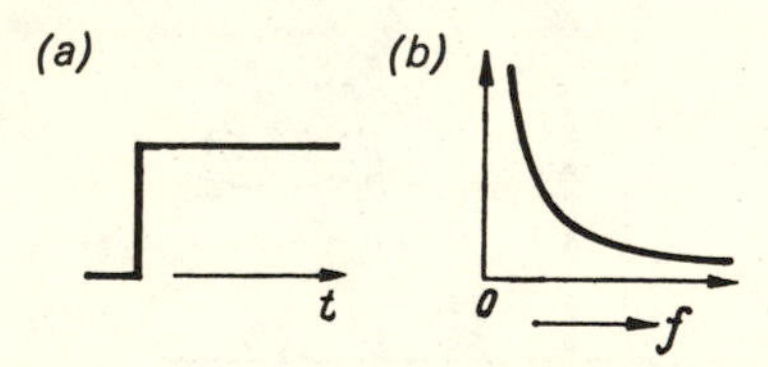

FIG. 22. (*a*) Onset process of direct current as a square wave of infinite period; (*b*) envelope of the spectrum.

An infinitely short onset time does not occur in nature, and hence not in the realm of sound. *Natura non facit saltus.* If the onset of a vibration circuit occurs during a certain time T (Fig. 23), correspondingly fewer vibrations will occur in the sound spectrum immediately after the onset, as will be more fully explained below (p. 42).

The most nearly soundless "switch," or, acoustically speaking, noiseless onset of sound, occurs when the very many lines of a spectrum—the partials—are reduced to a minimum. This requirement will be met by onset characteristics with a time curve

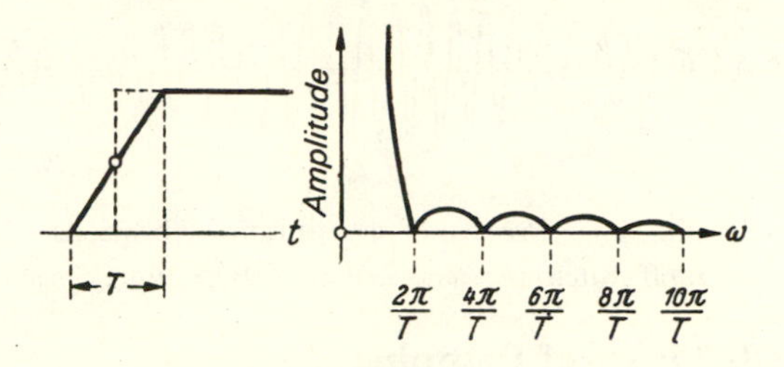

FIG. 23. Onset process of duration T with the envelope of the spectrum.

represented by an ever constant transition, as in Figure 24 (logistic curve) (106).

All onset events which occur in the world of musical sound lie within the physical bounds of these two extreme cases: the inconstant function of sudden change and the ever steady transition. A number of different onset curves with the corresponding

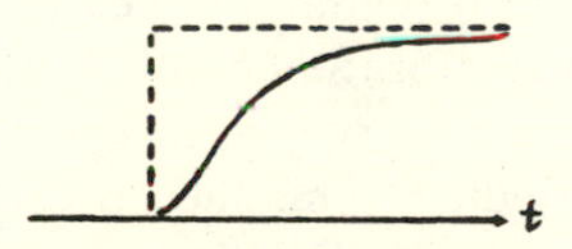

FIG. 24. Steady transition between two levels of vibration.

partial spectra have been computed by K. W. Wagner and collected in a table (106).

These laws for the onset of sound are valid also for changes in dynamics, for here too we have a transition effect from a condition 1 to a condition 2, e.g., from *piano* to *forte*. They are also valid for changes in pitch and tone color. We speak then in general of a transition function (Fig. 25), which is mathematically computed exactly like the onset event.

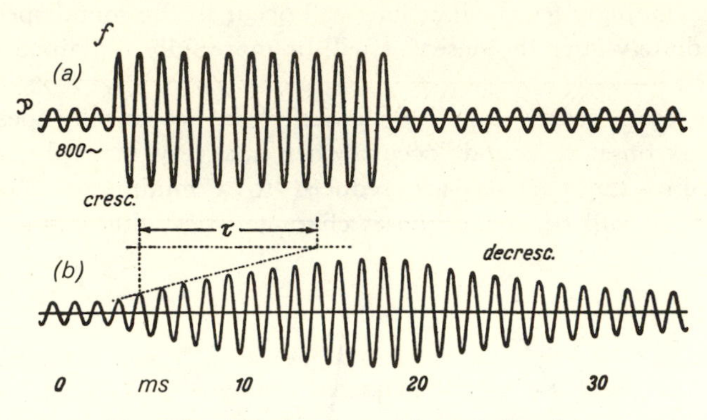

FIG. 25. Process of transition of vibrations.
(*a*) on/off switch; (*b*) *crescendo/decrescendo;* τ: time constant.

3. The Influence of Damping

First we have observed the behavior of stationary vibrations: they exist only when they have infinite duration. If this requirement is not satisfied, and the vibration event begins at a certain time, then at the first instant, in addition to the desired vibrations of the frequency f_0 there is an infinite number of other vibrations of all frequencies.

In discussing the above, we said very little about the system which generated the vibration, the sound emitter. These systems have a certain damping effect, that is, they partly absorb and partly emit the energy introduced to them. A struck piano string vibrates less and less after the first vibration, until after complete exhaustion of the energy it comes to rest. Such damped vibrations are not, strictly speaking, periodic. This then is the general practical case which alone will interest us in further observation of sound emission.

Figure 26 shows how the decay in amplitude of the sine wave (I) can be represented through the superposition of several undamped sine waves of just slightly differing lengths of period (frequency). The principle is represented here with only three sine waves of neighboring frequencies f_1, f_2, f_3 (II). The addition of ordinate values at a point in time A results in the summation value B. This can also be shown experimentally: A damped tuning fork emitting a damped train of vibrations of an average frequency of 100 cps, as in curve I, will excite three other tuning forks set up at appropriate distances with frequencies of 99, 100 and 101 cps. A

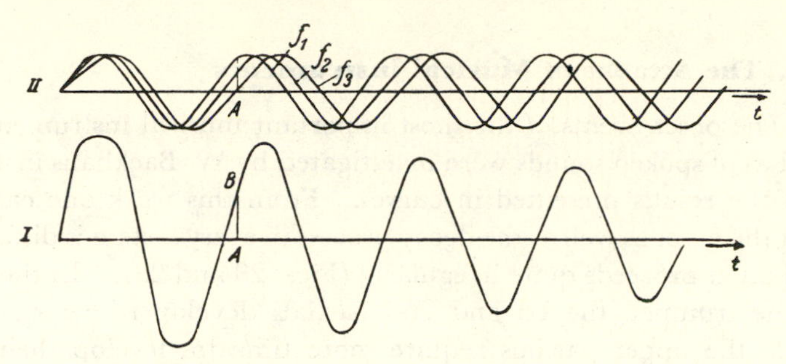

FIG. 26. (I) Amplitude change of a sine wave; (II) seen as the superposition of several undamped vibrations (f_1, f_2, f_3).

certain damping of the vibration generator is desirable for the stability of the sound generation. Besides, the ear, as a damped receiving system, would not be able to distinguish between single, discrete vibrations out of a train of vibrations of a certain bandwidth.

In a similar way, when we have only two neighboring frequencies the phenomenon of beats can be explained (Fig. 27). Now it

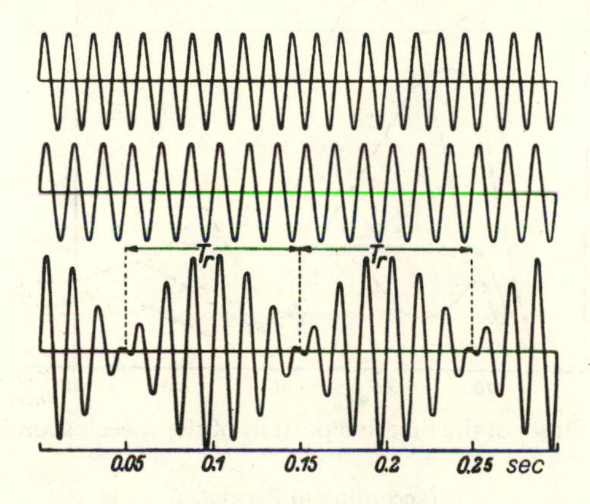

FIG. 27. Origin of beats from two sine waves by superposition.

can be seen that every change in the sine wave curve, in amplitude or in frequency, leads to a division into several vibrations of different frequencies, a condition generally referred to as "transient"; with attentive listening this can often be recognized as a momentary change in sound, often as noise. The degree of change required for it to be audible has been experimentally established (p. 112 ff.).

4. The Attacks of Musical Instruments

The onset events of the most important musical instruments as well as of spoken sounds were investigated by W. Backhaus in 1931, and the results presented in curves. From this work one can see that the onset as well as the decay process in every system is different, and often proceeds quite irregularly (Figs. 28 and 29). In the case of the trumpet, the 1st and 2nd partials develop rather quickly, while the upper partials require more time to develop their full energy; in the case of the violin, the first two partials develop more slowly than the upper partials. This is the reason why the trumpet sounds more clearly defined, with more fundamental, than the violin. The constantly changing relationship of the intensities of the partials occurring during the build-up of the sound is responsible for the real content of the sound in the sense of a development of color modulation.

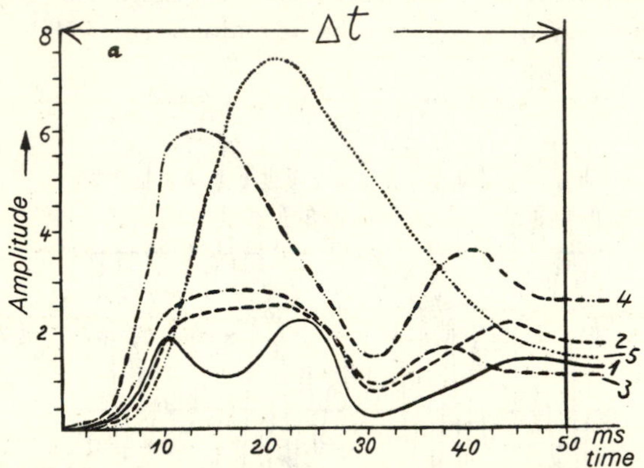

Fig. 28. Onset of the first five partials of the speech sound complex "dah" (according to Backhaus).

We know from experience that the ear does not perceive such very irregular fluctuations at the onset. The explanation for this lies in the fact that the inertia of the hearing mechanism causes the fluctuations over a period of time to be integrated, as will be explained in detail below. Thus one speaks of a time constant of the ear, approximately 50 milliseconds.

Figure 28 shows for comparison the onset of the spoken sound "dah." It comes closer to the sound of a trumpet, in the sense of a plosive, than, e.g. to that of a violin. The stoppage (tongue against the upper teeth)* brings a delayed suddenly opened aperture for the air stream, with sharp increase of air pressure, causing the typical consonant sound of "d."

The individuality of the onset process of each instrument—described above as irregular—comes from the complicated mechanical construction of the individual acoustical bodies. Even at the initial formation of a sound there are acoustical transients existing

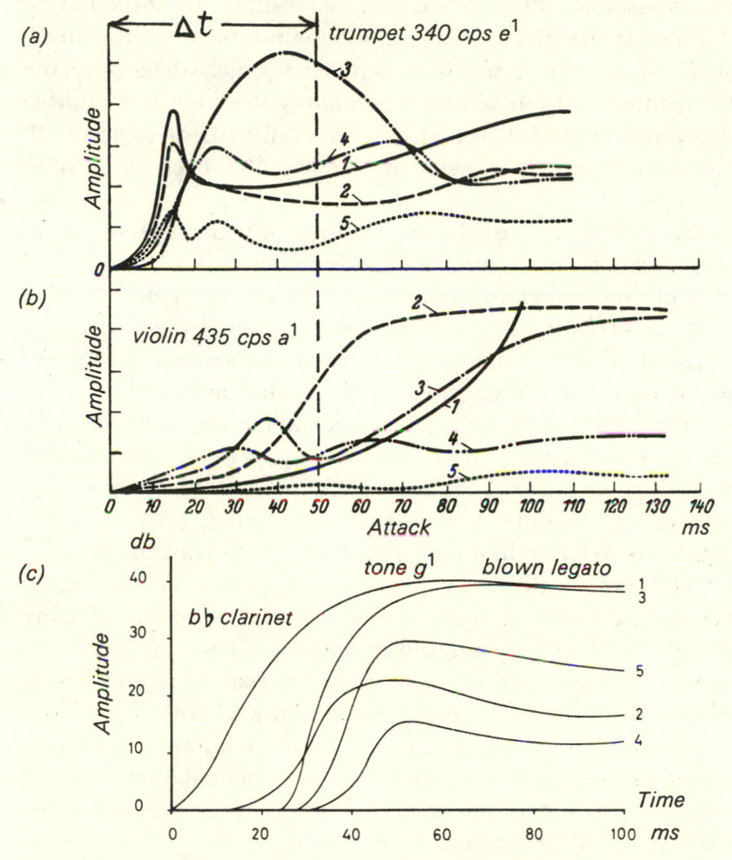

FIG. 29. Onset of the first five partials of musical instruments (*a* and *b*, according to Meyer and Buchmann; *c*, to Winckel).

* In pronouncing English "d" the tongue is against the alveolar ridge.

in coupled systems, such as string and body, reed and bore of the mouthpiece, the various cavities of the human articulatory organs, etc.

In addition, there is not only a simple switch process, analogous to that of electrical alternating current, but there is a gradual formation or deformation of sound, dependent upon the playing technique of the instrument, the pressure of the fingers on the strings, the pressure of the bow, the plucking of the taut string, the air pressure in the case of wind instruments, and finally the mechanism of the instrument, etc. It is desirable that the onset and decay processes of instruments be as variable as possible and not too brief, for they determine the quality of the sound to a great extent. Speech is exclusively a series of transients which determine the semantic content. In general this can apply to all kinds of audible assertions, also to music. This has been recently made clearer with the help of information theory, which has also been applied to music.

Carl Stumpf (99) has shown that an instrumental sound of constant pitch and intensity loses its character to a certain extent if one "cuts off," that is, renders inaudible, the typical attack. Music which has been recorded on tape can be cut up with scissors. The cut piece of tape, when played back, is the same as an instantaneous onset, the sound effect of which can be computed.

With such experiments Stumpf has been able to establish through a number of observers, that such "amputated" instrumental sounds lead to unexpected confusion. A tuning fork, for example, was mistaken for a flute, a trumpet for a cornet, an oboe for a clarinet, a cello for a bassoon; but even more contrasting tone colors could not be differentiated, such as cornet and violin, or French horn and flute. It must be observed that the sounds of many instruments (bells, piano, plucked instruments) have no stationary condition, but consist entirely of transients (decay after once being excited). If one were to generate synthetically a known sound of a bell from the spectrum partials, it would be a constant sound, and therefore quite unlike the original. But the bell-character would immediately be present if the sound complex were brought to sound in the dynamics of the original time units.

In musical sounds the characteristic overtone spectrum (formant) and the onset and decay transients are of equal importance. This is unfortunately overlooked in recent works on musical aesthetics, which again and again deal only with the stationary part of a sound through its overtone structure.

Physically, a regular dependence of the onset time on the frequency spectrum can be established; Küpfmüller has presented this in his *Systemtheorie* (45). Two cases are particularly interesting to us here: namely, the sound source and the frequency curves, which show one or more formant peaks—or else the opposite: holes in the spectrum (Fig. 30) (16). In the first case, with the presence of a hump, the stationary end condition of a sound will be

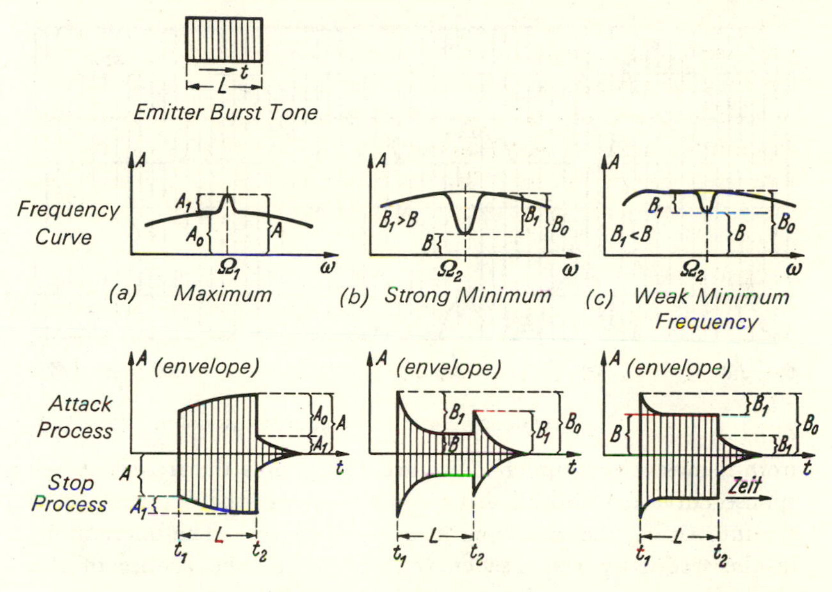

FIG. 30. Relationship between the spectrum and the onset process: (*a*) Hump in the frequency curve; (*b*) large trough; (*c*) small trough.

reached slowly and without "over-shooting"; the smaller the hump the more slowly the end condition will be reached (Fig. 30, case *a*). A corresponding decay is recognized in the switch-off of the tone at time t_2. If, however, there is a hole in the relatively smooth frequency curve, the maximum sound pressure B_0 will be reached immediately upon switching on the sound source (case *b*); this pressure then slowly decays to the constant value B. In switching off there is an over-shooting at B_1, at which point the decay begins. In the case of a much smaller hole, this is no longer the case (case *c*).

We can judge the *behavior of the onset transients* from that of the acoustical radiation of the instruments. The *frequency response curve* is of special significance here, as it is for a loudspeaker. We can discover this, for example, in the case of a violin sound box in the following way: we can excite the violin body by using the vibrations of a tone generator for all frequencies in the range of 80 to 16,000 cps, and at some distance from the violin body, using a measuring microphone, we can determine the strength with which each tone arrives. They do not arrive in equal strength, as one should expect

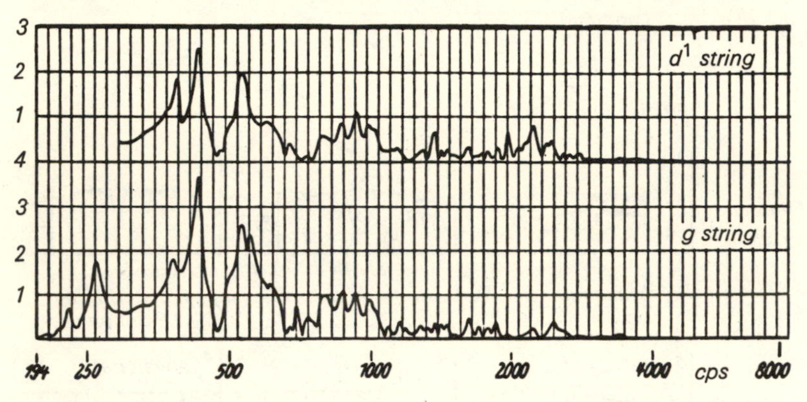

FIG. 31. Frequency curve of a Stradivarius violin.

from a musical instrument. Figure 31 presents the frequency response curve of a Stradivarius violin, showing the dependence of the intensity of the vibrations at all frequencies. The fluctuations in the frequency response curve account for the timbre of the instrument—of course, in connection with the onset transients. For the purpose of comparison, Figure 32 shows the frequency response curve of a good loudspeaker, which, as a transmitter of all sound events, should respond completely neutrally, and should therefore theoretically have a straight-line frequency response. A further example of a frequency response curve is seen in Figure 2.

One requires of the soundboard of a good piano that it have no special resonance points, so that all tones can be emitted with equal strength. In practice this is only partially possible. For this reason, the sound intensity of struck piano tones is very unequal even with excellent pianists (p. 47). The human vocal tract (throat and mouth) is adjustable, for instance, to the resonance of certain frequency areas as a coupled resonance system, whereby the vowel formants originate (p. 13).* To sing a scale with equal loudness for every tone requires careful training of the singer. This can be checked with the volume meter of a tape recorder.

A test was used to check the reproducibility of the single sound sources (116). Experienced players of the different instruments in the orchestra were asked to play a scale over the full range of the instrument for the purpose of producing the same level for every tone controlling intensity only by the ear.† Some examples for

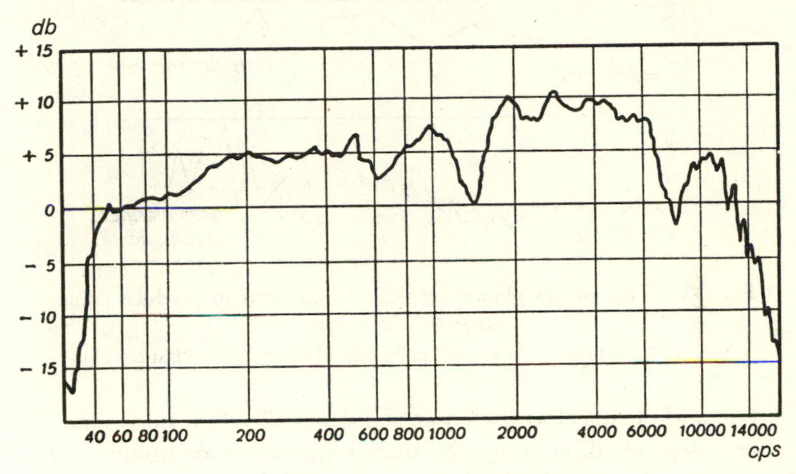

Fig. 32. Frequency curve of a loudspeaker.

these measurements, taken in a little studio fitted with thick carpets and curtains, are shown in Figure 33. The curves begin with the lowest possible tone of the instrument and go on in steps of a full tone of the diatonic scale, the last example in a chromatic scale. The deviations from a reference level are very irregular and have a width of 10 db or 20 sones or more. This depends partially on the characteristics of the instrument because the repetition of each test shows nearly the same irregularities as the first one. For the other instruments in the orchestra, similar tendencies in the sound-pressure curves were obtained. By the use of the instruments in a choir, one gets a balance of the level. Nevertheless, due to the skillfulness of the conductor and the sensitivity of his ear, the

* An exact physical explanation is given by H. K. Dunn, G. Fant, etc. See bibliography.

† The author is indebted to the Cleveland Symphony Orchestra and its conductor, Dr. George Szell, for making the mentioned experiments possible.

orchestra is able to produce a sound level which, at every repetition, varies on the average only by ±1 db. The players were, of course, conscious of the technical difficulties related to maintaining a constant level.

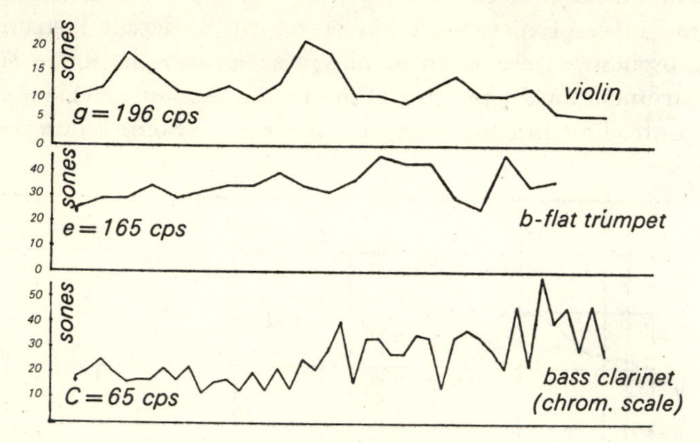

Fig. 33. Tone scales played by skilled musicians to produce equal subjective loudness.
Violin and trumpet in diatonic scale, bass clarinet in chromatic scale.

The audible effect of the onset transients of a damped resonance system depends directly on the magnitude of the damping. This magnitude determines in which frequency area f_1 to $f_2 = \Delta f$ the system can vibrate. These limits are defined by the half width value of the resonance curve measured in half the height of the curve. The onset time τ can be computed as

$$\tau = \frac{1}{f_2 - f_1} = \frac{1}{\Delta f}$$

In the theoretical case of an undamped system, the frequency area Δf would be infinitely small, the onset time infinitely long. From Figure 6 we see that the resonance curve *B* is more greatly damped than curve *A*, and therefore the system, corresponding to curve *B*, responds to a greater area of frequencies f_1 to f_2, and the onset time τ for this greater Δf is shorter.

Such strongly damped systems are absolutely desirable for the production of music and speech, for only when the onset time is short enough is it possible for different sounds to follow sufficiently quickly one after the other. The simple tubing of a flute has less damping than the body of a violin, and, consequently, has a correspondingly longer onset time, about which more will be said below (p. 42). Because of this one can play faster passages on a violin than on a flute.

Many attempts are being made today to revise the classical notation of music in order to take into consideration the subtle structure of sound. A three-dimensional representation (68, 70) indicates which factors are necessary for this. The individual sound

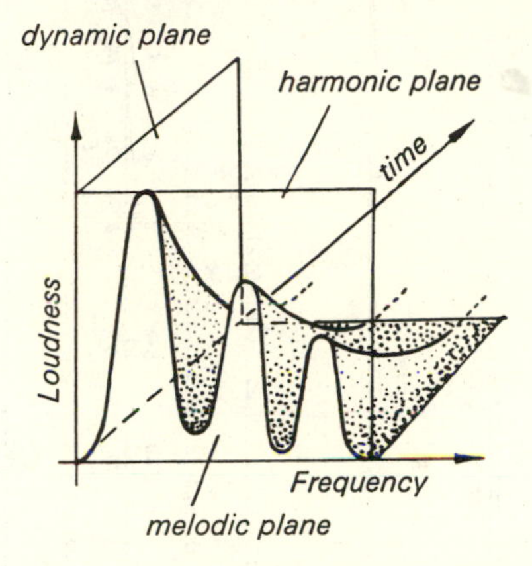

Fig. 34. Microstructure of a sound with three formants in a three-dimensional plot.

is represented in dynamic, melodic and harmonic planes (Fig. 34), from which the onset and decay processes are expressed on the time coordinate and the overtone structure on the frequency coordinate, while the changes of the overtone spectrum in time are seen in the plane formed by the frequency and the time.

Today, known and new sounds and noises are generated on tape as phonomontages. Figure 35 presents an example of how a composer with the help of his own symbolic representation can give instructions for the generation of an electronic composition. It was surprising to discover from the compositional attempts to date that besides the above-mentioned factors of effecting sound there are still further phenomena of a psychoacoustical nature which operate to affect the experiential power of music from quite a different angle. Also during experiments in speech segmenting we have learned how the inner connection of the transient processes contains functioning elements of information, about which we still

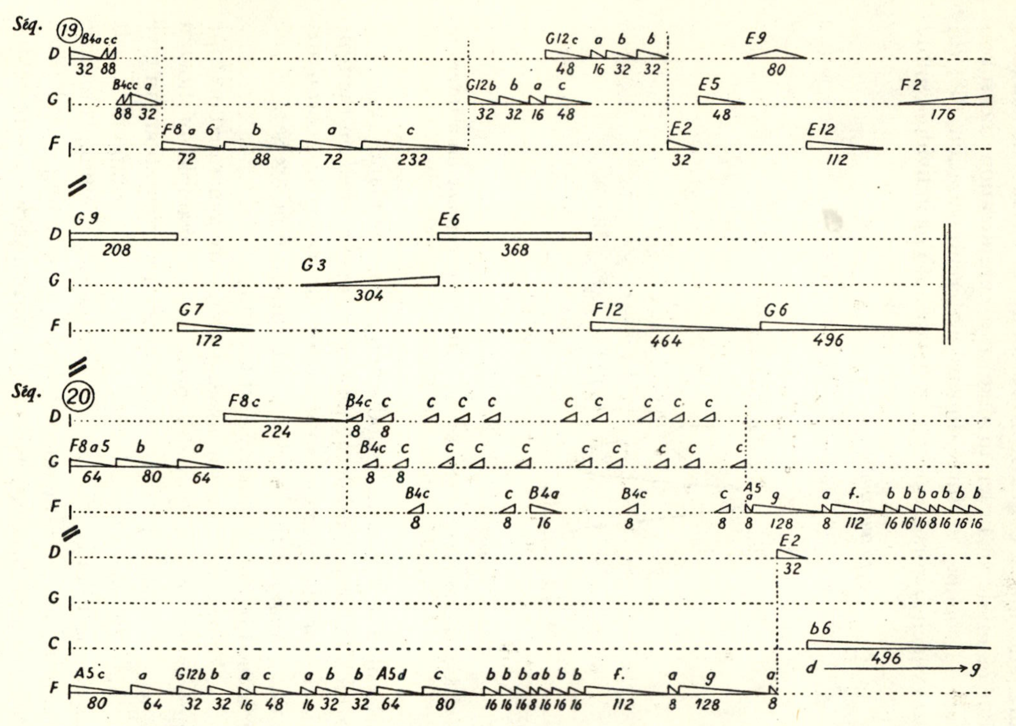

Fig. 35. Extract from the notation of an electronic composition by Olivier Messiaen.

know very little (18). This problem is dealt with in the next sections.

5. The Notation of Sounds

The discussions above have shown us that musical sound has a very dynamic life. The classical representation by means of a note head seems insufficient for the needs of contemporary composers, who are already beginning to sense the complexity of the individual sound, and who intentionally make use of its inner generative powers for the flow of a series of tones. The extent to which the precise observance of dynamic and agogic refinements in performance is a concern of modern composers is shown by the many printed indications in their scores, as, e.g., in Figure 36. The principle of a

FIG. 36. Luigi Nono, *Varianti*, 1957.

dynamic development over an extended stretch as a special mark of Romantic music has been discarded in favor of the use of loudness signs as significant elements of single sounds. It should be recalled

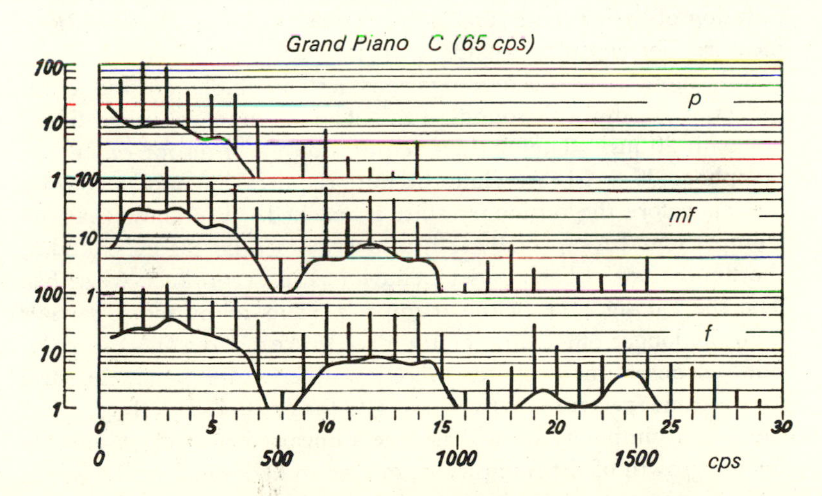

FIG. 37. Spectra of the piano touch curves.
Under the spectra: touch noise.

that the tone color changes with loudness (e.g., see Fig. 37), inasmuch as the sound effect with greater loudness seems to be noisier and can even stimulate shock if unprepared, as in the leap from *pp* to *ff*. This means an energy load on the perception organs, all the more if such a dynamic change is repeated so that the ear is required constantly to re-adapt. Imagine such changes in light intensity in a film presentation. A one-hour performance would lead to nervous exhaustion. An interpreter is thus required to select slow tempi for music constructed in this way.

6. Variable Onset Times

Having given the quantitative information on the physical magnitude of the onset times (switch-on times) in general, it is of interest to give some figures for the practical onset times of speech and musical sounds. The onset time is very short for vowels, about 6 milliseconds (ms), independent of the vowel color, which is presumably characterized by the stationary behavior of the sound. The duration of a vowel in most Western languages lies between 0.04 and 0.20 sec. In an analogous way we can discuss the voiced semivowels l, m, n and r. In contrast, the real consonants are largely to be regarded as a transitional phenomenon between two vowels; consequently, they often represent transients for the whole duration of their sound (Fig. 38). Times of 5 to 15 ms have been measured for explosive sounds (p, t, k).

Still more detailed investigations are available on the onset processes of musical sounds (3). The trumpet is exceedingly fast with 20 ms. The clarinet requires 50 to 70 ms for onset, the saxophone 36 to 40 ms, while the flute requires 200 and sometimes 300 ms before the stationary state is reached. We know that the trumpet sound is especially rich in overtones, whereas the sound of the flute is not. Their onset behavior is in accordance with this. The short onset time of the trumpet permits many more partials than the longer onset time of the flute. We see this in Figure 29. For each onset time T chosen there is a different frequency measure for the same spectrum corresponding to the relation $T = 1/\Delta f$. For the sake of simplicity we will assume a linear course as in Figure 23 for the growth of the intensity from zero to maximum at the onset, which in the case of musical instruments can be quite varied. It is then seen that on sounding a tone most of the energy content is concentrated in the very low tones from nearly 0 to the frequency $f = 1/T$. That means that the vowels at onset, that is, for a duration of from 5 to 10 ms, bring forth essentially frequencies below about 150 cps; in the case of musical instruments, the trumpet theoretically has its energy concentration below 50 cps and the flute below 4 cps.

As it is known that frequencies below 20 cps are not perceived as tone character, but as single beats, so it is that in the case of the

FIG. 38. Sonagram of the words "unusual pictures."

flute there is no audible onset distortion. In the case of the trumpet we can list a frequency band of from 20 to 50 cps as onset noise.

If we investigate the onset behavior of the partial components of a sound individually, we notice that they develop in different ways, as exemplified in Figure 29 for the trumpet. It is interesting that an intensity maximum for all partials occurs at 14 ms, at which time decay sets in. This behavior appears to be characteristic of the trumpet. In the case of the violin the relationship is quite different. The partials build up more slowly (Fig. 29), and the build-up time becomes greater with increasing partials. Similarly, the sound of the clarinet begins first with a fundamental, and then the overtones appear gradually, yielding a softer sound.

A valuable supplement is to be found in investigations of *organ pipes* (102). The sound of the reed pipes is built up and switched off quickly and precisely within the first to second period; both the beginning and the end are accompanied by a click (e.g., trumpet register). In the case of flute pipes the beginning of the sound is slower and less sure. After an initial noise, the harmonics appear

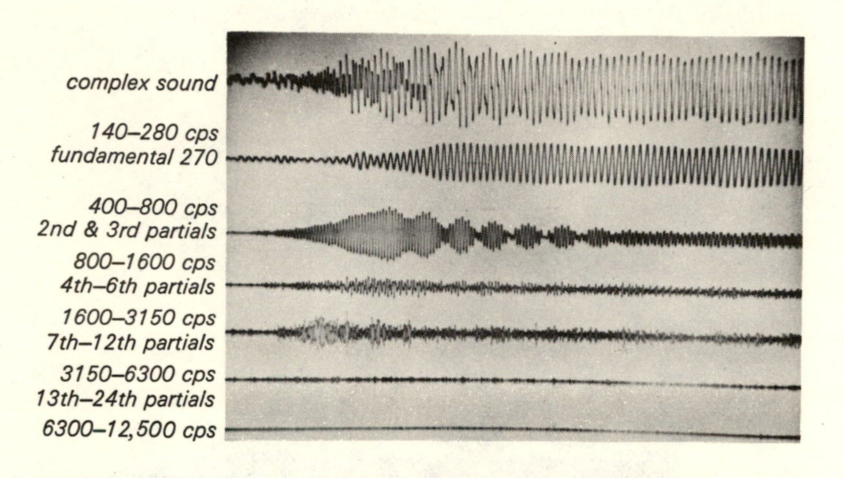

FIG. 39. Onset of a labial pipe of an organ.

Recording time 300 ms. Excitation of an inharmonic partial (interval of about a ninth; cf. 3rd line vibration from above) by an edge tone near the 2nd resonance forming beats with the octave. This partial "anticipates" the others (according to J. Meyer, PTB, Germany).

and then, finally, the fundamental. For one principle of the organ of the Eosander-Kapelle in Berlin-Charlottenburg, the considerable onset time of 0.6 sec has been measured—as opposed to reed registers with 0.1 sec. Through special control of the onset event curious musical effects can be generated. For example, the "Lieblich Gedackt" register of the same organ possesses a harmonic, actually an "inharmonic," the $5\frac{1}{2}$ partial, which appears for a short time and dies away, while the remaining harmonics build up. Figure 39 shows how the single partial areas of such an organ sound can be separated from one another by an electric filter and made visible on an oscilloscope. Even some inharmonic partials can appear disconnectedly during the onset. This characteristic response emphasizes the above-mentioned register in the case of rapid passages in polyphonic pieces, and thereby contributes to a clearer definition of the different voices. The peculiarity of the onset sounds of Baroque organs is employed particularly advantageously in the works of J. S. Bach.

It remains now to be shown how individual playing technique on an instrument or the attack of a singer can basically influence the timbre of the sound. The representation of the build-up in the case of a violin sound in Figure 29, for example, must be changed from this standpoint, inasmuch as the onset is to a great extent dependent upon strongly changing bow pressure (58). When the bow is first applied, the pressure at the first instant is too small to excite the fundamental vibration. Therefore, only the higher partials appear immediately; gradually, with increasing pressure of the bow, the lower harmonics and finally the fundamental appear. It is therefore part of the art of the violinist to attack with sufficient bow pressure, and on the other hand a necessary quality of a good violin that it respond quickly. Only then does the timbre appear free from the blurring and unclarity of imperfect reproduction. The fewer overtones in the sound spectrum, the easier it is to achieve this. Since Helmholtz we have known that it is not important to generate many overtones of great volume, but rather to achieve the right energy distribution within the spectrum through the distribution of the concentration of the formants (see also p. 13).

The onset behavior of combined organ sounds is further dependent on the type of air chamber employed. The smaller pipes of organs with a single air chamber speak first, afterwards the deeper pipes which are supplied with air from the same chamber. The amplitude develops much more constantly when there is an air chamber for each register. Therefore (102), the organ with an air chamber for each register is especially suitable for slow playing, chorales, etc., while the former is particularly adapted to polyphonic music as a result of the more precise attacks.

From the example of the organ, commonly thought of among laymen as a generator of stationary sound, we see the importance of the effect of the onset event on the character of musical reproduction, and how it can even influence the style of a musical work.

In this connection it is worth mentioning that Max Reger generally played on organs with an air chamber for each register.

The degree to which the complex onset phenomena provide the secret of beautiful organ sounds has been made very clear today by the many attempts in building electronic organs. It has been recognized that merely switching on a simple electric circuit is insufficient to recreate synthetically the inner life of the organ sound.

We can observe extremely varied attacks in the *human voice*. Deficient physiological preparedness (which may include a lack of mental concentration) while singing prevents the resonant areas of the physiological system from vibrating completely at the very beginning. Only after the vocal cords have begun to vibrate does the thus unprepared singer correct the adjustment of the inner resonant cavities until the optimal resonance is achieved. The aesthetic effect is unsatisfactory if the head tone with high overtones begins alone and the chest tone then follows, while the reverse is equally unsatisfactory. When the attack is correct, that is, when the inner adjustment is correct, a colorful healthy tone, which can develop naturally from *piano*, is immediately in evidence. Corrections made after the attack cause inconstancies in the "ideal" condition, which means one approaching an exponential function, and are perceived as adjustments and therefore as sound distortions.

The same observations are valid for *wind instruments*. In this case an external mouthpiece must be excited in addition to the human resonance system of mouth and throat cavities. Wind players still do not generally recognize the importance of physiological adjustments apart from the correct embouchure. A wind player who was first a singer will understand this best, as he will have learned to hear through an inner muscular feeling.

Pianoforte tone reacts in an opposite way to the blown tone. It quickly achieves its maximum intensity and then slowly decays. The well-known experiment of playing a piano recording backwards demonstrates the extent to which the process of sound decay is the reverse of the onset process: one hears a tone rather like that of a harmonium, that is to say, one of a totally different sound character, although the curve is the same. This is not the place to discuss in detail the extremely complicated string vibrations, which O. Vierling has investigated in detail (105). The important thing to notice is that the striking of the string also brings forth inharmonic components. The unsteadiness of the stroke has the effect of exciting all resonant frequencies of the soundboard. The soundboard, however, has been so constructed that it reacts equally loudly as nearly as possible, to the pitch of each string. Naturally, the inharmonic components decay relatively more quickly.

A definitive explanation of the influence of the piano hammer stroke on the timbre as a whole has never been given. We know simply that the tone color is altered by a change in the force of the stroke, as shown by the spectrum in Figure 37. A light stroke results in a softer sound with fewer overtones, a medium stroke makes the sound harsher, while a strong stroke gives the sound a bright, almost wind instrument-like timbre. Even so, this explanation is not complete. A different physiological playing posture of the pianist results in secondary effects yielding a varying timbre. One might point out an American experiment at the University of Pennsylvania in which the attack of a well-known pianist was compared with the attack of a weight falling on the key. The recorded vibration curves were identical (E).

The decay of piano sounds is just as interesting as their onset. The very different damping of the individual overtones of the different strings is particularly noticeable. It can even happen that a swelling in intensity occurs during the decay of a partial; this can be traced back to oscillation in the energy of the multicoupled resonant system of the piano (105). When the pedal is employed, a reduction of 45 db in 60 sec has been measured (1:170), which would correspond to an 80-sec reverberation.

It is especially interesting to consider the *behavior of the conductor* from the standpoint of transients. Energetic conducting with short, clear cues yields a timbre which, although laden with energy, is harsh. An overly forceful cue causes each instrument to begin too suddenly, which in the case of a *tutti* can lead to noisy onset distortion. According to the above discussion, a sudden attack causes a spreading of the acoustical spectrum, and now the partial components are multiplied by the number of instruments, while in addition, subjective combination tones are formed between the partial components. An imprecise attack by several instruments together extends the onset process by jerks, and, because of the resulting blurring, it is no longer characteristic of the desired sound.

The timbre is quite different in the case of a conductor who is able to catch up the attack or, in other words, to take up the attack with the elasticity of his bodily movements—a conductor who immediately gives an "anti-beat" to the beat. In that case one can scarcely speak of a beat-technique, it is more nearly a zig-zag motion; observed microscopically, it is a combination of small movements. The attack prepared for in this way is acoustically soft, even considered as a physical onset transient, but is still capable of an increase in tension. This method of conducting was more typical of Wilhelm Furtwängler than of any other conductor. He was capable also of adjusting the attack to the acoustical behavior of the concert hall, or of its onset response. It was possible for the author to arrive at these observations through a comparison of recordings made in connection with the testing out of a remodelled concert hall (113, 114). Analogously, it is also possible to derive characteristics of other conductors, which this author has done. Thus the entire personality of the conductor as a psychosomatic whole is mirrored in his handling of the baton.

The onset and other transients in the generation of sounds determine to a great extent the form of the sound of a musical work. We are thinking primarily of the color quality of the sound with its accompanying noise, but one must also consider that the *course which the tempo takes* in the performance of a musical work depends on the time constants of the onset and decay transients of the sound source.

It takes every tone a certain amount of time before it can begin to sound after its onset click, that is to say, before the stationary condition can be approximated and the real sound formed. Rapid passages, depending on the time-beating technique of the conductor or performer and also depending on the acoustical behavior of the room, can be increased in speed only within relative bounds. We will have occasion below to speak of the limits set by the ear. On the other hand, the decay of one sound sometimes dovetails into the next sound, leading to desirable or undesirable harmonic or inharmonic combinations, which can greatly influence the musical effect.

Not even the *dynamics* can be formed completely arbitrarily, for—as we have seen in the piano example—the onset transients and the resulting tone color are dependent upon them. A convincing demonstration of this is playing the radio alternately loud and soft.

There are therefore close interrelationships between the musical elements of harmony, dynamics, tone color and tempo. Similarly, we must consider the attack of sound physically and physiologically, as well as the behavior of the instrument, the player, the room and finally the ear. We will return to this below when discussing Figure 110. Only after consideration of all of these psycho-acoustically important factors can a musical performance achieve a successful effect.

We have taken for granted the further factor of the psychical readiness of the listener, which in turn is dependent on environmental factors and must reach a stable condition. The surprising failure of a performance, or the failure of a composition which appears good on paper, can frequently be explained by totally exterior circumstances, e.g., that an excellent violinist unexpectedly is forced to play an instrument—perhaps better than his own—that he is not used to playing.

7. The Indeterminacy of Tone Perception

In looking back over the above observations we see that the physically exact pitch of a sine tone cannot be fixed, for it would require infinite duration in order to become a discrete vibration of frequency f. A vibration occurring in time always forms a frequency band Δf, which we can determine for the time span Δt during which the vibration lasts. If, on the other hand, one attempts during the time span Δt of the train of vibrations to fix a moment of time one will fail here too, for reducing the time span Δt to a moment leads to an infinitely broad frequency band, as we have seen in the discussion of the onset transients.

Time and frequency are bound together through the laws of nature; we have already seen this formulated as

$$\Delta t \cdot \Delta f = 1$$

The more exactly we fix one magnitude, the more inexact is the determination of the other. For example, the more precisely we wish to perceive the pitch or frequency, the less precise will be the perception of the *microduration.* Thus we must settle for a compromise in selecting a good relationship between Δt and Δf. There is a limit to the accuracy of measurements, a limit which one only reaches with a certain inaccuracy. Abbé ran up against this problem while attempting to determine raster fineness under the microscope; and today this relationship plays a great rôle in atomic physics under the name of Heisenberg's principle of uncertainty, which operates with the two concepts of *place* and *impulse*. From the above formula we can conclude, for example, that a tone with sine wave characteristics at 100 cps with the duration of one second can only be determined as closely as 1 per cent. This law of uncertainty cannot be applied to non-sine wave vibrations, as shown by Licklider (53). In the application to sound spectra, the uncertainty of frequency, for example, means only that short limited waves have many frequency components. A wave form with few frequency components, that is, with narrow band width, can be fixed with accuracy.

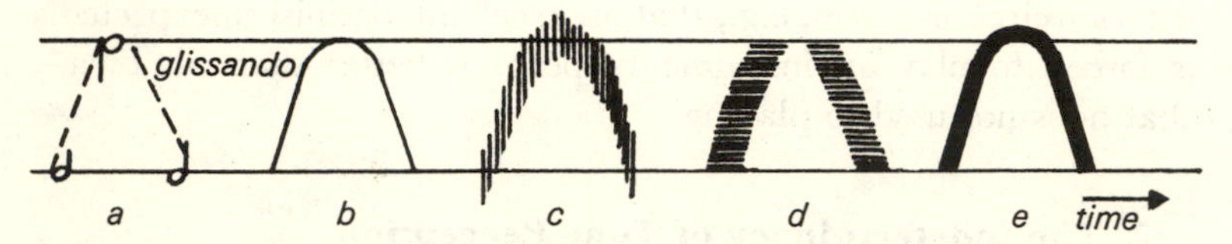

FIG. 40. The uncertainty of a glissando tone in frequency and time.

Therefore it is not possible to establish the pitch at every instant of a glissando, as we can see from Figure 40*b*. Exact pitch is not revealed in any point of time, nor can any exact point of time be linked with a determined frequency. Suppose we think of the glissando as being dissected into a chain of tones which are sounded rapidly after one another. Let us imagine further a hearing organism which is adjusted to a very high frequency selectivity; from this mechanism we will get only a very blurred impression of the precise duration of individual tones (Fig. 40*d*). If on the other hand the hearing mechanism is constructed to recognize precisely the amount of time involved in an event, it would get only a rough impression of the individual frequencies (Fig. 40*c*). Nature, therefore, has constructed the human ear as a compromise between perfect accuracy of time and frequency perception for optimum hearing of speech and music (Fig. 40*e*), a point to which we will return below.

A measuring system such as the human ear requires a certain damping of the receiver-resonator, in order that it can react quickly enough to the frequency changes to be indicated. An undamped resonator would require an infinitely long onset time. We can deduce information about the damping of the resonators of the inner ear by observing the fastest change in frequency which the ear can still perceive. In the case of a logarithmic damping decrement of roughly 0.25 in an average frequency area of from 200 to 1000 cps, band widths of the ear resonators of about 25 cps were arrived at. This is what determines the width of the band in Figure 40*e*, and therewith also the characteristic of human music and speech perception.

It is conceivable that some animals have different damping values in their ear resonators, with which they would then perceive music and speech differently; in the case of greater damping more rapid series of tones would be possible which, to be sure, would then sound more noisy (large Δf), and the reverse would be the case with less damping. It is also possible that such abnormal hearing occurs in pathological cases of human beings. It is well known that for some people music contains more noise than for others.

The formation of the sound phenomena described above is easily demonstrated experimentally by means of a tape recorder. First one inscribes on the tape a sine wave of constant amplitude, which can be recorded from an electrical sound generator, or if that is not available, one may inscribe an evenly blown flute tone; then one cuts more and more segments out of the tape little by little, until finally there are only one or two vibrations left on the tape.* In listening to this, after each cut one will notice how the sound becomes more noisy as its duration becomes smaller. One can study onset transients of different kinds by cutting tapes at different angles.

The indeterminacy of time and frequency have the effect of an audible smearing of an exactly defined note value for a certain time of the attack. How long is this time?

8. Time Constant of Sound Perception

In order to arrive at a reasonable unit of measurement, let us observe the real course of the transients progressing from a low to a high volume (Fig. 25, bottom). If we draw a tangent to the peak amplitude of the increasing vibrations and set the maximum of the undamped vibration as limit, we can derive a unit of measure for the time constant. According to recent investigations by M. Joos, this value has been given for hearing at about 50 ms, equal to $\frac{1}{20}$

* The narrow strips of tape can be pasted onto blank tape so that they can be played back on the tape recorder.

sec, and has been labelled by him "perception time smear" (38). Because of the principle of uncertainty we cannot arrive at the precise ending of the sound smear; however, we can assume practically that it is no longer perceivable after two or three time constants, that is, after 100–150 ms. In any case it is particularly noticeable for the duration of one time constant. During this time 63 per cent of the highest possible amplitude change is reached. The connection of change in amplitude with frequency distortion can also be accounted for by the fact that the change of a single sine wave into a following larger sine wave cannot be represented by a pure sine wave. The distortion of the sine wave is thus responsible for the formation of the color-changing partials (Fig. 27).

The value of 50 ms as the time constant is surprisingly large. Let us not forget that in rapid conversational speech a phoneme also requires precisely this span of time. And when we remember further that the smearing is still noticeable throughout 2–3 time constants, then we can imagine of what importance this is for music reproduction. After all, we can produce on a piano a maximum of 12 strokes per second—just as on a typewriter—so that the individual sounds have a duration of about 80 ms (37).* The sound development is therefore broken off before the onset transient event has been completed. Stable sound is simply not formed.

The smearing of the ultimately desired tone quality does not generally mean that the tone color during the 50 ms after the beginning of the sound is made completely unrecognizable. The extent to which that is true depends in each case on the spectral combination of the sound complex present and, further, on the extent to which the spectrum of the previous sound differs from the following sound. In special cases simple sounds of higher frequencies can be recognized after a switch-on time of 10 ms, so long as the following 40 ms are silent, so that the ear can reconstruct the acoustic fragment. We arrive in this way at the time constant of "sound recognition ability," which will be discussed below. An important result for the performance of music is that the average tone color of a sound series varies with the speed of execution of the series.

One might remark in this connection that the time constant of

* Still shorter note values, such as grace notes, are no longer independent sound events for us, but rather are specific appendices of the onset event of the entire sound.

50 ms has a far-reaching significance; for example, in the study of environment (*Umweltforschung*) it represents the so-called *human factor.* According to Karl Ernst von Baer (1860), this is the time that we require to become conscious of one of our sensory impressions. Although this constant has been given in the most varied areas in a rough approximation between $\frac{1}{10}$ and $\frac{1}{20}$ sec, recently the human factor has been more accurately set at about $\frac{1}{18}$ sec = nearly 55.0 ms (13). In music these border values are of great significance. They correspond very well with the values obtained from the damping decrement of the ear resonators (p. 50f.). It is also shown in this relationship that as frequency increases, the time constant takes on smaller values. Measurements of the ability to distinguish pitch have given, in the case of a pitch at 100 cps, the value of 50 ms; at 1000 cps, 20 ms; and at 4000 cps, 14 ms (20).

It is not possible to say with certainty today just what connections of a higher biological order obtain. In order better to view the relationships, it should be mentioned that the biological factor of other organisms is determined through their own time constants. This ordering in the world of Siamese fighting fish is $\frac{1}{30}$ sec. That means that changes of less than $\frac{1}{30}$ of a second cannot be recognized by the fish. One can see in this short time constant that this creature is faster-living than man. A snail, on the other hand, has the considerably larger time constant of $\frac{1}{4}$ sec = 250 ms, five times that of man, indicating its relatively greater slothfulness. An average value of 3–4 ms has been established for a number of insects, which explains their fast reactions. Thus it is clear that all these beings must hear speech and music differently, in correspondingly distorted forms from the human standpoint.

For times shorter than the time constant there is the law of stimulation quantity: power × time = constant (Bunsen–Roscoe law). Accordingly it would be possible that a strong stimulation of 10 ms, for example, would seem just as long as a weaker stimulation of, say, 35 ms because of the equal total energy content.

In the following chapters we will show how the critical length of echo effect, reverberation time, stereophonic effect and so forth is dependent upon that universal constant which is established as the integration constant of the ear. It can be demonstrated that sound pressure impulses occurring in intervals greater than 50 ms no longer blend together but are received separately: echo effects, as well as fundamental tones which lose their tone character below

20 cps. Their noisiness can be explained by the fact that in the spectrum the partials lie closer together than 20 cps—the threshold where a stable sound becomes noise.

On the other hand, we note that when two short tones of different pitch are struck one right after the other too rapidly, at least during the 50-ms period, they cannot be perceived surely in the right order (metathesis). It actually occurs that when consciously listening to a piece of music it is sometimes impossible to determine the correct sequence of melody tones in a theme when it is stated very quickly. This phenomenon is known in physiology as *gegenwartsdichte*. Very rapid playing adds a colorful flickering or rustling, which certainly seems intentional in the *Jeux d'eau* of Liszt and Ravel.

Similar statements can also be made for the tempo of speech. The duration of the sound units of speech, and thus the separation into vowels and consonants, is determined by the integration time of the ear, 50 ms. If this time constant were considerably greater, articulation would be very unclear, as if in a room with a long reverberation time. Conversely, in the case of a shorter time constant, articulation would become overly crisp, as can be heard when playing a tape recording of speech or music at twice the normal speed. This gives us an idea of how animals with greater time constants than man hear speech and music. The corresponding case of a smaller time constant can be demonstrated by playing the tape more slowly.

These remarks should prove sufficient to show how in the human world principles of music aesthetics are determined by a physiological constant: the boundaries of the tone spectrum, the speed with which tones follow each other, the evaluation of the tone color, the "space" perception of sounds and the ability to determine the direction of a sound source.

Let us apply the observations made thus far to musical events, which physically are to be regarded always as a series and superposition of trains of vibrations limited in time. As will be explained in more detail in the final chapter, single trains of vibrations of stationary character never last longer than $\frac{1}{10}$ sec in a practical musical performance without the appearance of a new sound event, or the superposition of a new one upon the old. As a result of this limitation to a time interval t, the discrete acoustical vibration corresponding to the note head is widened to a whole band of vibrations of frequency Δf, whose width we can compute according to the formula of K. Küpfmüller:

$$\Delta f = \frac{2}{t}$$

Here we see another connection with the principle of uncertainty, in that the factor 2 arises simply through the choice of the coordinates. For limited sound events of $\frac{1}{10}$ sec there results a band width of simultaneously emitted vibrations

$$\Delta f = \frac{2}{1/10} = 20 \text{ cps}$$

Shorter sound events result in a correspondingly wider frequency band, for which reason music played faster than normal sounds noisier, hence more dizzying. In any case, in music we are always concerned with a band width of at least 20 cps—regardless of the pitch. The band width of 20 cps means for the pitch *a* a quarter tone above and below, for *G* in the suboctave it means a full tone above and below (Fig. 41). From this we see that the ability to

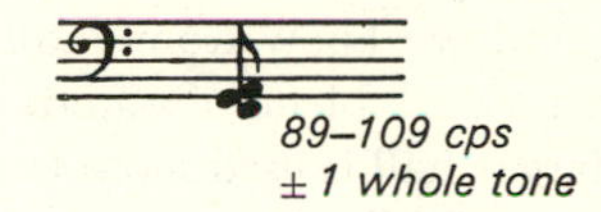

FIG. 41. The musical effect of the expansion of the frequency band of tones of $\frac{1}{10}$ sec duration.

distinguish between pitches of short duration improves as the pitch rises.

Just as we have seen that a sound event in music cannot occur more rapidly than about 20 in a second, we have also discovered that the boundary necessary for an elementary sound event to become musically effective is from $\frac{1}{20}$ to $\frac{1}{10}$ sec, or from 50 to 100 ms (Fig. 111).

9. Fullness of Tone

Transients have the further function in music of giving the sound the quality of fullness and spatial perspective.

Theories and experiments have shown that one cannot locate the sound of a sine tone generator through distance and direction estimation (L). In this case, too, it is an interrupted series of tones rather than a continuous tone which aids the ear in the location of the source. In technology the effects of clicks or shots are employed for acoustical location, since they produce a very wide frequency spectrum. The advantage of this we shall see immediately.

According to G. v. Békésy, the location of a sound source, especially the determination of distance, does not depend in general only upon the transients, but also particularly upon the presence of low tones, especially in the area between 0 and 500 cps (A). Mathematical investigation reveals that high tones alone do not permit an estimation of the distance. Sound sources with predominantly low frequency transients are perceived as near. If these are not present, the sound source is perceived as farther away, as is also revealed by oscillograms taken with microphones placed at different distances from clicks or shots. At the same time one gets the impression of a bright tone color. In the reverse situation, the low frequency events give the effect of a full sound, which is so meaningful for performances of music. G. B. Shaw once criticized orchestral performances of the BBC, saying that there was too little bass, and we know from radio that the taste of the public leans toward a dark blend of sound. This accounts also for the fact that as a larger hall is used more bass instruments must be employed or, stated physically, the low transients must be emphasized in relationship to the stationary part of the sound. Concert managers and conductors too easily forget that music of the classic period was written for smaller halls, and that therefore a performance in a larger hall without an alteration in the make-up of the orchestra cannot yield the same effect.

If there are too few deep tones in a sound picture, the ear will compensate to some extent by creating the difference tone from two simultaneously sounding tones—or from two or more partials of a tone. In the case of spectral partials of sounds, the fundamental is formed anew subjectively in this way, and it sounds considerably sharper than would a sine tone of the same pitch. This phenomenon—as long as we are not dealing with a combination tone—is known under the name "residuum" (86); it happens only for sounds of less than 40 db. The fundamental is an important factor in the three-dimensional contour of sound.

This will be discussed more thoroughly in Chapter IV. If the simultaneous beginning tones are inharmonic with respect to each other, the difference tones are particularly noticeable. This explains one of the effects of dissonant chords, while the other effect lies in the closeness of the spectrum (p. 6), and therewith replaces the portion of noise necessary for a piece of music.

Let us remember that an extremely harsh sound attack which is equivalent to a switching transient emits a very broad and close frequency spectrum and therefore gives a noisy impression, that of a click. For the aesthetic perception of musical sounds it is important that the deeper tones be sufficiently received and that on the other hand the highest tones in the spectrum be more or less damped; otherwise rapid passages would simply sound like noise, and stationary sounds would possess too much dissonance above the 7th overtone. This then contains a certain hint for the construction of musical instruments and for the training of the human voice, which are analogous electrically to a low band pass filter. But these statements are in contradiction to high fidelity technique.

Chapter IV

THE CONCEPT OF SPACE

1. Criteria for a Music Room

A sound event which reaches our ear contains all of the criteria for the room; this can be registered on an oscillogram and interpreted, as we shall demonstrate below. This includes the characteristics of the room itself, and further the direction and distance of the sound source—in short, all information which even a blind person can gain when he claps his hands in a room and listens for the reverberation.

A room, just as a musical instrument, has specific onset events. Sounds from a source do not only come directly to the ear, but come also secondarily resulting from reflections off the walls, whereby the total impression of the sound is deformed by the acoustical characteristics of the room. The onset event of the sound attack is extended for the ear at least until all of the reflected sound has reached the ear, the sounds which cover the greatest distance arriving last. The resultant onset events, which either form a reverberation onset situation which continues the formation of the attack or a reverberation extension of the decay of the sound, can be extended under certain physical conditions to yield an echo effect, intolerable for music. If the walls reflect very well, an echo effect can be noticeable when the difference in time between the sound and its reflection is more than 50 ms, and the intervening time is not occupied with other reflections. A time span of 50 ms corresponds to a distance of 57 ft.* Concert halls for large orchestras are

* Path $S = c \cdot t$, where c is the speed of sound at 1115 ft/sec.

generally longer than 57 ft, so that echo presents a serious problem but one that can be solved by acoustical engineers and architects.

A sound impulse thus draws secondary impulses after it through the reflections back and forth from wall to wall, first from the nearest wall and then more and more from all sides, so that the concentration of impulses increases while the energy decreases (Fig. 42). The recording of a reverberation curve is made in the

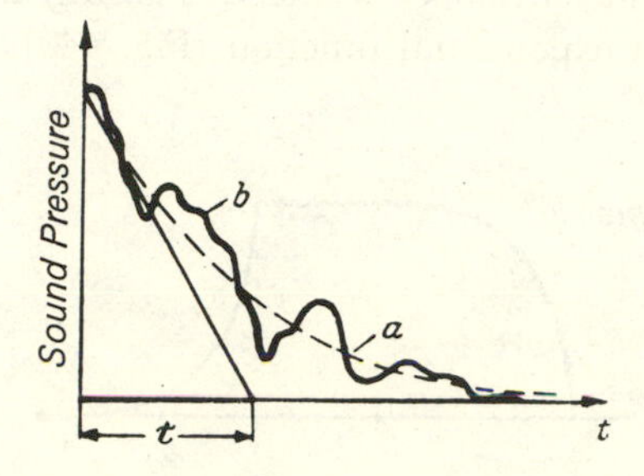

FIG. 42. Decay of the sound energy.
(a) Theoretical (exponential curve); (b) real (according to measurements).
τ = time constant employed for the computation of the decay.

following way: a banging sound is generated in the test room, and the reflections of the sound impulses are picked up by a microphone, amplified and projected on the screen of a cathode-ray tube. In the left-hand oscillogram in Figure 43 there is a fairly even

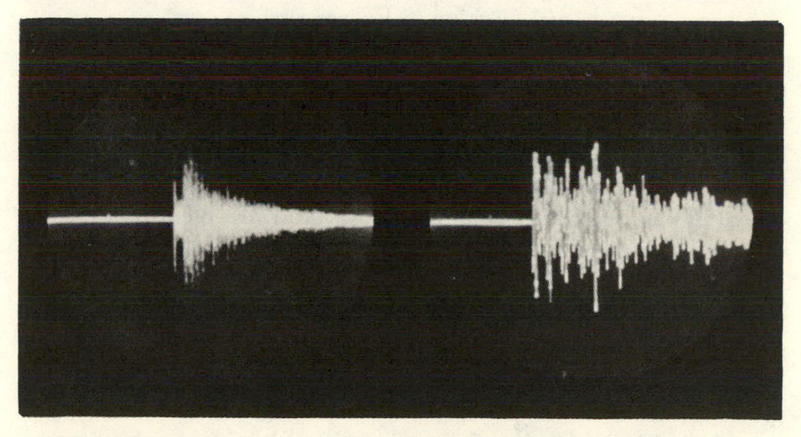

FIG. 43. Impulse diagram of the wall reflections of a room (for example, a concert hall).

distribution of the impulses coming statistically from all sides in a chronological series; in the right-hand oscillogram, however, the series of impulses is very irregular even in amplitude. Whenever two strong impulses follow one another with too great a lapse of time, there is danger of the formation of echo. This method of measuring can also be used to test models of concert halls before the actual construction in order to study the acoustical characteristics of the room. The envelope of the decreasing impulse peaks is approximately an exponential function (Fig. 44*a*). In Figure 44*b*

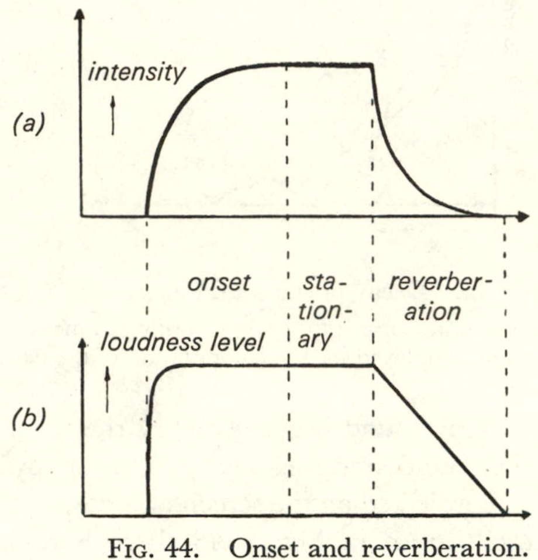

FIG. 44. Onset and reverberation.
(*a*) Intensity; (*b*) loudness level.

the curve has been reckoned logarithmically in order to correspond to the logarithmic characteristics of ear sensitivity. A blind man senses a closed room by perceiving the envelope as a decay in loudness. The impulses return from the enclosure following quickly one after the other, as explained above, and are blended through the inertia of the ear.

The decreasing energy briefly finds compensation in an increasing density of impulses, until the falling-off tendency becomes too great. This maintaining of the intensity has been called by G. Boré *Verweilzeit* (lingering time) (8), and by Békésy *Präsenzzeit* (presence time); it has an optimal duration of about 50 ms, as an empirical value. During this period the original sounds seem to be prolonged and amplified, and therefore "clearer," while the secondary impulses which arrive later are considered a disturbance, because they distort or smear clarity through this overlong reverberation effect.

These relationships were clearly presented by H. Niese (Fig. 45). In a beginning area of reverberation many strong reflections are desirable, and they are called useful sound (*Nutzschall*) since they contribute to the definition of the sound structure. If the integration time of the ear, the threshold of smear (about 50 ms), is exceeded, the further reflections should decay according to an exponential curve. The impulses that exceed this curve are called disturbing sound (*Störschall*) since they cause an effect like an echo.

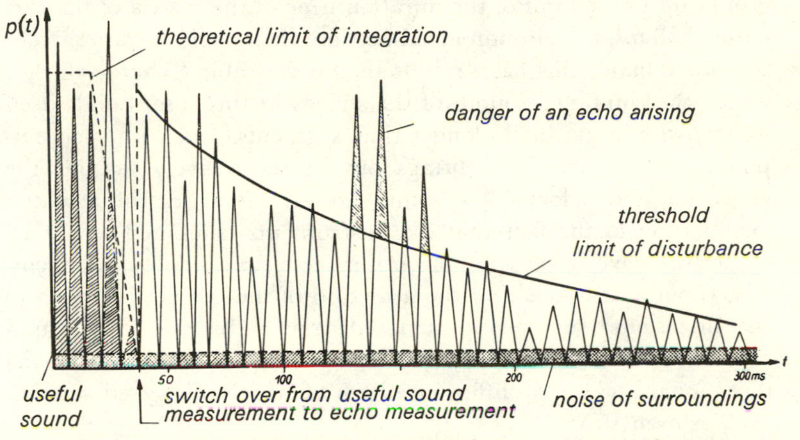

FIG. 45. Diagram of sound reflections in an architectural space. Generator: sound impulse.

Thus R. Thiele (100) forms an expression for "definition," placing the intensity during the initial 50 ms in proportion to the whole decay of intensity:

$$D = \frac{\int_0^{50\text{ ms}} I(t)\,dt}{\int_0^{\infty} I(t)\,dt}$$

Thus a new concept has originated which is of great interest to musicians also. The English language had already employed the expression "definition" as a requirement for satisfactory reproduction of music in a concert hall. In the past few years many halls have been built which possess good definition, a transparency of the score, but at the cost of some of the fullness of sound and flow of musical events. Perhaps this suits the music of our time, as it fits polyphonic composition, but it is not suitable for the sound requirements of the Romantic period. The dependence of music on the architectural acoustics is so great that it can be maintained that the polyphonic form of composition could only have developed when suitable buildings were created with the corresponding transparency of sound.

Acoustically "dry" rooms, which frequently have exaggerated definition, are also unpleasant for musicians because the slightest inexactness in playing is immediately noticeable. Nature provides only one protection: inexact attacks in ensemble, e.g., of a string group, up to the limit of the duration time of the inertia of the ear, namely 50 ms, remain unnoticed because they blend together. On the other hand, this even results in the desirable "choral effect," where the combined sound, of the strings in this case, builds itself up stepwise in gradually longer time segments, and with such emphasis on the transients brings out in more lively fashion the character and color of the sound body. Here lies the essential explanation of the fact that like instruments combined in an orchestra can exhibit much more brilliance than a single instrument. Also, one must notice that the attention of the listener is drawn to the first sound source and is not distracted by the other sound sources that enter directly afterward (31). This explains the directional and space-diffusion effects of widely dispersed sources, such as orchestras.

To summarize, we can see from this that it is not necessary that attacks in an orchestra occur with absolute precision, since here again there is a permissible range of delayed starts, which even serves a useful function when the limit of 50 ms is observed. Delayed attacks between instruments of different tone color certainly should be more critically perceived by the ear. Sudden attacks at once result in a broad, concentrated spectrum, so that the hearing function over a broad frequency area is blocked for the following attacks, as explained above, even though they may be of another timbre.

The conductor who gives a hard beat thus generates a sound structure of greater exactitude, regardless of whether the attacks of the individual instruments are together.

The blocking effect in our hearing, which by the way has also been called an echo-block, has been determined quantitatively by H. Haas (31) through a loudspeaker experiment. When in a large assembly room at a great distance from the lecturer a loudspeaker operates by transmitting the speech with a delay of up to 50 ms from the original sound, the sounds of the loudspeaker are not noticed, and the attention of the listener is focussed on the lecturer. Surprisingly enough, this is also true when the loudspeaker is generating a sound about 10 db greater than that of the lecturer. The physiological explanation for this is that during the 50 ms after the sound impulse the cortical cell groups cannot be stimulated, and an impulse arriving during this time span will not be perceived (41).

The influence of the architectural space on the quality of sound is also quite noticeable during the room onset (*Anhall*). Let us take for example a room with thick sound absorbing material on the whole ceiling or a wall. Here the sound absorption of one or more surfaces reduces the number of reflections and the build-up of the sound is thus more quickly completed, as shown in Figure 46 by the dotted-line curve. Such sound lacks the roundness and the three-dimensionality of the "symphonic" sound of some of the best halls (see p. 70). Recently there have been attempts to measure this rise time because of its great significance.

Just how long it can be in relation to the reverberation, can be clearly presented, according to Skudrzyk (93), by giving the time ordinate of the build-up curve in per cent of the reverberation (Fig. 44). One can see here for example that with a reverberation time of 2 sec (concert hall) the build-up is essentially completed after approximately $\frac{1}{2}$ sec, while in a theater with 1 sec reverberation time the build-up takes $\frac{1}{4}$ sec, or a little more than the time of a syllable in a word (about 200 ms). From experience we have learned that symphonic music has optimal effect in rooms with a 2-sec reverberation time, and consequently we would expect frequently to encounter sound elements having a duration of about 500 ms in this music.

If we plot the general curve of sound build-up (Fig. 46) by taking the absolute values of loudness sensation in the sone scale, we find that the best rounding-off of the build-up, which we perceive as three-dimensionality of sound, is achieved at an average loudness level of about 90 phons—roughly *forte*—whereas the curves become flatter as the loudness declines. On the other hand, there would be a too rapid ascent of the curve in the case of increased loudness (more than 100 phons); this causes difficulties for the adaptation of hearing, and also a burdening of the nerves.

In the case of halls with very long reverberation times, e.g., churches with more than 7 sec, the build-up process is drawn out in time so long that the curve again appears very flat. This corresponds to our hearing experience, thus supporting what we said above, that a sound intoned at *forte* in a room with about a 2-sec reverberation time yields the best fullness of tone.

An analogy can be found in looking at pictures. Single elements in drawing and painting receive their three-dimensionality not

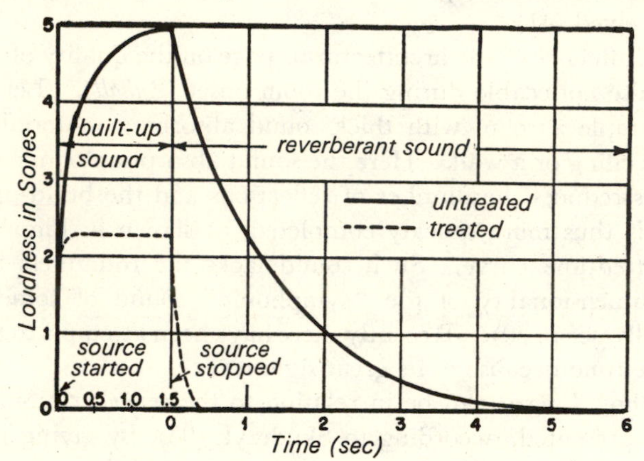

FIG. 46. Loudness of build-up of reverberant sound before and after acoustical treatment of a highly reflective room (according to H. J. Sabine).

through sharpness of contour but through a kind of "swell" of shading toward the outlines of the elements. This is particularly well done in paintings by Lovis Corinth and Emil Nolde.

For musical aesthetics one must not overlook the fact that the definition of the musical texture is reduced with increasing reverberation; this is also true in the mentioned analogy with painting. One stylistic feature of classical music actually consists in allowing clarity to be somewhat obscured in favor of harmonic interweaving in successive series. A sort of inner sound building and sound development is in the foreground at the price of the rational composing of tones without structure. It was especially in the Viennese classical period that the concert hall was thought of as a formative element.

In practice there is a rule of thumb for the formation of echo, which is that in larger auditoriums loudspeakers should not be spaced further than 57 ft apart from the source, corresponding to the distance sound travels in 50 ms, for otherwise the sound impulses from the source and the loudspeaker would no longer be blended together by the listener.*

The architect's art, or the skill of the acoustical engineer, consists in constructing the hall in such a way that resonant peaks will be avoided, for in these frequency areas there would be preferred sound utterances and the remaining tonal areas would be more strongly damped. Just as a violin body is so constructed that as far as possible a relatively equal sound volume is emitted in all positions, so it is desirable for a music room to have a similar equality in its frequency response, that is, for all audible tones to be heard with approximately equal volume.

Reverberation is avoided completely in a room where all the walls and ceiling are covered with nearly 100 per cent sound-absorbing material. This corresponds to sound reproduction in the open air, since there is no sound reflection there, either. Such damped rooms are used for measuring sound sources where the influence of the room must be eliminated (acoustically dead room). By employing wall coverings that have certain acoustical reflection characteristics, and with planning and computation, the reverberation time in new buildings can be brought to a musically desirable value.

M. Philippot reports the lowest limit of reverberation time for music reproduction as 0.8 sec. If during an orchestral performance a sound to which our attention has been directed suddenly ceases, 0.8 sec is required—as one can establish through tests—until another instrument in the score is noticed (75). This time span is the *adagio* tempo, which can apparently be assimilated especially easily by the untrained ear, and which provides the least strain on the memory. From the example of the diminished seventh chord of the *Coriolan Overture* (Fig. 105) we will see that after the sudden attack of the chord there is at first absolute stillness until, after approximately 0.8 sec, reverberation is perceivable. Philippot

* It is interesting that the same time constant is valid for optical perception. Two television or movie images which appear less than $\frac{1}{18}$ sec apart, blend with one another, while in the case of a slower series first a flickering is seen and finally the images will be perceived individually.

considers this a psychological effect inasmuch as attention is diverted from the primary sound source to sound coming from other directions and forming reverberation (attention split). In a room having less than 0.8 sec reverberation time, no flow or slurring within the tone series would be noticeable beyond the original sound emission, since a shorter sound emission is not noticeable.

2. Optimal Reverberation Time for Music Reproduction

In order to make a proper judgment about the reverberation times measured in practice, it is necessary to have a standard value. According to the definition of W. C. Sabine (1900), reverberation time is the time that elapses from the instant the sound source is switched off until the sound energy decays to one millionth of its original potency (60 db). The dependence of the reverberation time T on the room properties was empirically found by Sabine to be

$$T = 0.16 \frac{V}{A}$$

where V is the inside volume of the room in cubic meters and A the total sound absorption of the room, which is essentially that of all the room's surfaces. Thus the reverberation is dependent on the size of the room and the characteristics of the materials of ceiling and walls, but not on the shape of the room.

The reverberation time actually perceived by the ear is, to be sure, usually shorter. We are not able to follow the sound decay up to the threshold of perception because the sound melts beforehand into the omnipresent background noise in the room. Experience tells us that this is never less than 30 phons, is usually 40 phons, and can even be 50 phons. If on the other hand we are dealing with *p* or *pp* places in the music, the usable loudness is only 10 to 20 db above the background noise level. That which audibly decays is thus only a third of the value employed in Sabine's definition. In spite of this, Sabine's standard value for reverberation will be used here, as it has been introduced internationally.

The question arises, which are the most appropriate reverberation times for music? In the case of symphony concerts in large, fully occupied concert halls, values of 1.4 sec to 2.2 sec, with reference to an average frequency range of 500 to 1000 cps, are suggested.

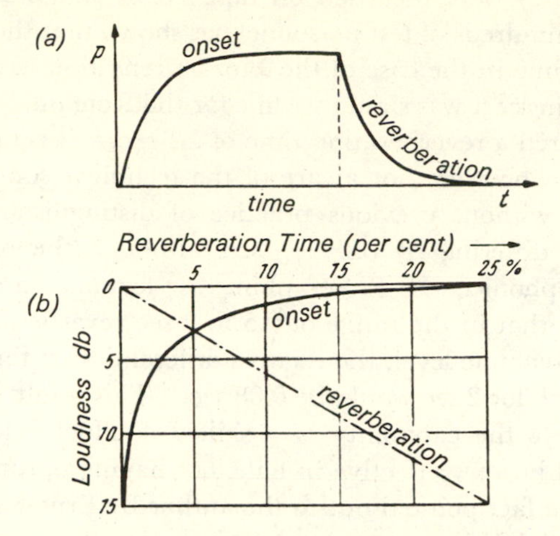

FIG. 47. Time plot of the sound pressure after switching a sound source on and off.

(*a*) Linear measure; (*b*) logarithmic measure.

We found that 2 sec was an especially good build-up time for symphony concerts. In churches reverberation times of 7 sec and even more have been ascertained. The most appropriate value depends upon the type of music being performed. Chamber music demands the low value of 1.4 sec. Conversely, it might be maintained that musical styles and genres have developed in accordance with the architectural settings available in given periods. In the enormous stone Gothic cathedrals with an inner volume up to 250,000 m^3 = 32,699 yd^3 (Cologne Cathedral), a monophonic Gregorian chant in slow tempo was best adapted, while in the Leipzig Thomaskirche (18,000 m^3 = 23,543 yd^3) with only 1.9 sec reverberation time when fully occupied, polyphonic compositions with faster tempi could flourish, and—more extremely—in the modern "dry" studio atonality and twelve-tone music, as well as pointillistic music, are perfectly at home.

The sensitivity of music listeners to the smallest differences in reverberation time has been ascertained by a cooperative experiment of German radio stations (46). In a number of studios with differing reverberation times a typical work of the classic period (Mozart's "Jupiter" Symphony), a work from the Romantic era (Brahms' Fourth Symphony) and a piece of more recent music (Stravinsky's

Sacre du Printemps) were recorded on tape. The judgment of the recordings by hundreds of test personnel has shown that the optimal reverberation time in the case of the Mozart rendition was 1.5 sec. With the Stravinsky it was similar, while for the Romantic piece the majority preferred a reverberation time of 2.1 sec. It is interesting that musicians who were not aware of the technical context were easily capable without previous practice of distinguishing reverberation times differing by only $\frac{1}{10}$ sec, although the recordings were not stereophonic. H.-P. Seraphim (91) formulated a general law which says that in the range of 0.5 to 2 sec reverberation time and 40 phons dynamic level, the relative differentiation threshold is 4 per cent, which for 2 sec would be 0.08 sec. This result, however, does not exclude the capability of a skilled conductor to form a sound that will be very effective in halls not having optimal reverberation time, a fact pointed out to the author by Ernest Ansermet and Hermann Scherchen.

The careful observance of the reverberation time, unfortunately has not always led to the desired result of good musical effect in the construction of concert halls. Good effect depends on the evenness of the distribution of the sound in the room, and here the geometry of the room plays a certain rôle which is often underestimated.

Through curved ceilings, for example—this is analogous to optical laws—acoustical foci can be formed. In addition, shutter echoes can occur between parallel walls, leading to unequal distribution of the sound in a room, that is, inadequate sound diffusion.

In addition, the balance in sound emission of all the instrument groups of the orchestra is important; the manner of setting up the complicated orchestral apparatus is significant, involving perhaps raising individual sound groups by means of risers, so that the rear instruments are not covered by those in front. Where the sound emission of the brass instruments is too strong, the reverse may be advisable.

If the reverberation time is too long, this resembles the effect of a pedal, and a too drastic leveling of the sound energy of all groups occurs, which destroys the definition of the musical texture. The transients which are characteristic of speech and music are covered by reverberation.

It is therefore not only the reverberation time but also many other factors which are responsible for the optimal acoustics of rooms designed for musical performances.

In order to provide a further clarification of these problems, the author sent out a statistical questionnaire in 1948 to leading conductors all over the world, asking which concert halls were acoustically the best for performances by large symphony orchestras (113, 114). A summation of judgments resulted in the list of concert halls presented in Table I.

It is noteworthy that the halls listed in Table I are mostly rectangular—except the Teatro Colón—and are decorated in a neoclassic style, with columns and sculptures, coffered ceilings and so forth. Precisely these features contribute to a *diffusion* of sound and to an extensive scattering of the reflections from the walls. The creation of statistically dispersed reflection of impulses, so that the impulses do not appear compact, is of great significance in the acoustical perfection of the room. This accounts for the fact that the heavily ornamented rooms of the Baroque period are frequently extremely good for music. The present dilemma is that the modern architectural style, especially in cubic forms, is diametrically opposed to that which would provide good musical acoustics. (Recently an acoustically more adequate architectural style has been developing through cooperation between architects and acoustical engineers.) These observations have been reinforced by experiments with artificial reverberation.

Problems of architectural acoustics have been considered here in such detail in order to show what an important participant in a performing ensemble the architecture of the room is, and that the laws of musical fluctuations must be applied here just as consistently as in the construction of musical works and the manual execution of them.

Another example shows the extent to which musical effect is dependent on reverberation. The lower part of Figure 48*a* demonstrates that several tones in close succession begin and decay, e.g., spoken syllables approximately 2 sec apart (67). While one sound has just reached its maximum intensity, the next sound has already begun. It is strongly masked by the reverberation of the preceding sound, so that speech—like music—is blended into a general and neutral sound, which is difficult to understand (pedal effect). The upper drawn-out curve shows how the sound complex would continue to grow with increasing intensity; here we see that the sound growth and decay times are equally important. The actual development of intensity of this sound series results from the

TABLE I

The Best Concert Halls of the World*

Concert hall	Reverberation time (occupied) 500–1000 cps	Year of completion	Vol. (in m^3)	Seats	Dimensions, incl. stage (in m)	Orchestra area	Walls	Ceiling	Floor
Musikvereinssaal, Vienna	2 sec	1870	15,000	1650 + 300 standees	42 × 15 × 18	135 m^2	plaster; 20 wood doors; 38 glass windows; 1 balcony	plaster on spruce, coffered	wood
Teatro Colón, Buenos Aires	1.8 sec	1906	20,000	2500	33.5 × 29 31.5 height	theater stage	7 balconies	plaster on metal lath	wood
Concertgebouw, Amsterdam	2.0 sec	1887	19,000	2200	28 × 44 × 17.2	100 persons + choir	plaster; 1 balcony	plaster on reeds, coffered	wood
Symphony Hall, Boston	1.8 sec	1900	19,000	2631	48 × 22.5 × 19.5	150 m^2	plaster; 1 balcony	plaster on metal screen	concrete slab with wood

* Extracted from a list published by the author in *Baukunst und Werkform* ("Die besten Konzertsäle der Welt") 1955, pp. 751–753 (some data are altered).

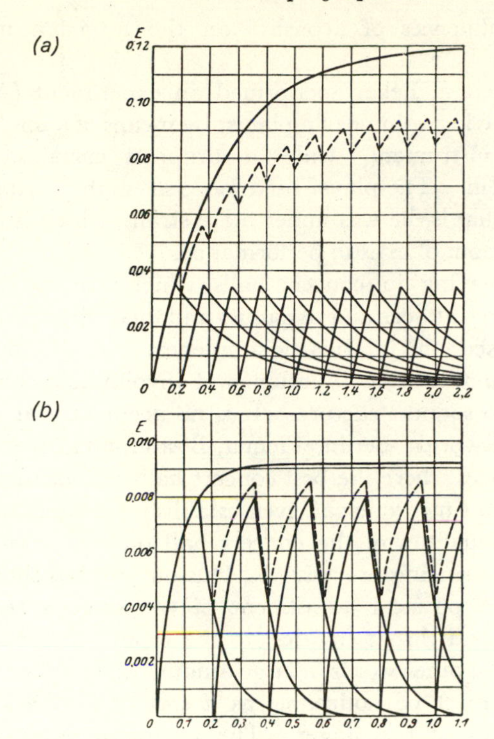

FIG. 48. Effect of reverberation time on music and speech.
(*a*) In the case of long reverberation time $t = 2$ sec; (*b*) in the case of shorter reverberation time $t = 0.4$ sec.

addition of superimposed sound components, as shown by the broken-line curve. The example in Figure 48*b* provides better conditions, in which the effect of a considerably diminished reverberation time, approximately 0.4 sec, is presented. The decay occurs so rapidly—sharp descent of the curve—that the new sound reveals itself clearly with very little masking by the previous sound. This negligible masking is in fact advantageous, because it constitutes a binding together of the speech sounds, and eases the flow. A further reduction of the reverberation time would not permit a sound to be formed by the room. Speech would be hacked and music "dry." Besides, it can be seen from the curves which musical tempo is best with regard to this blending of the tones.

Aside from the direct influence of the room on sound and sound flow, further influences of acoustics on the musician must be discussed.

In this connection, Békésy performed an experiment (A). He measured the playing volume of a pianist by means of a microphone built into the instrument, while he dropped curtains, greatly damping the room. The player used more strength playing in the damped room, that is, he was under more strain, which is undesirable for the freedom of artistic performance.

Conductors playing in concert halls which are too strongly damped must force attacks and dynamics and their physical power is unnecessarily taxed; this inhibits the suppleness of their baton stroke, which alone can transform intuition and involuntary conception immediately into sound structure. It is no accident that the best orchestras of the world—Berlin, Vienna, Boston, and others—have developed in places where the best concert halls are located.

In the case of a singer one can even establish the direct influence of the acoustic qualities of the concert hall on the physiological mechanism of the singing process. R. Husson (36) has shown that there must be a physical adjustment of the singer's resonance cavities (mouth and throat cavities) to the room—marked by the impedance—or, in other words, a large amount of acoustical energy from the room must be noticeable as a sort of feedback in the resonance cavities of the singer. The optimum value will be reached when the singer notices a certain resonance tickling on his palate, which results from this feedback effect and which leads to an adjustment of the vocal apparatus in the larynx through nervous excitation.

Stage fright of artists comes from the fact that they fear they will find no "resonance" with the audience; this can be traced back to a room which does not permit a "resounding" resonance. Many failures of great international musicians, as well as many failures of first performances of works that became favorites in the repertoire can be traced back to this cause.

3. The Directional Sense

As mentioned above, a blind person can ascertain the direction from which speech and music come. This is astounding considering that reflections from walls pour sound from all sides onto the listener, and that sometimes this secondary sound has considerable energy, especially when the walls are hard, polished and even. We have already seen from the experiment of Haas (p. 63) that with two equally loud loudspeakers the one that seems to be heard is that one whose sound impulse reaches the listener first. This is the one which is recognized as the sound source.

The blocking effect of the ear plays a rôle here (we spoke above about echo-blocking)—namely in that the attack of the second, more distant loudspeaker is not noticed. This occurs, however, only when the time span is greater than 3 ms. Up to this time limit the sound impulses coming from two loudspeakers and reaching the ear at slightly different times will be localized at some intermediate spot between the loudspeakers. The angle at which the ear receives the seemingly single sound depends upon the time difference between the two impulses. This is the process of phase stereophony, which exists alongside the technically preferred process of intensity stereophony, in which the ear locates the sound source through the volume differential between the two loudspeakers.

In summary: The ear discovers the direction of a sound source by the phase as well as by the volume differential of the impulses. A time lag between two impulses will occur when some natural sound falls on the ear sideways (Fig. 49), for the ear farther away from the

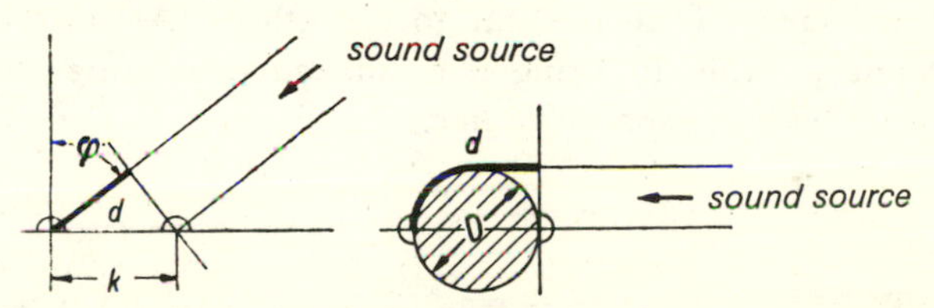

FIG. 49. Direction sensitivity in binaural hearing as a result of the phase influence.

sound source receives the impression later. Somewhere in the nervous system the two impulses received by the ears are compared, and from continual experience the appropriate decision about direction is made.

Let us assume that the sound arrives completely from the side, so that the distance d traveled from one ear over the head to the other ear is 21 cm (8.27 in) longer (Fig. 49). According to the formula already mentioned, the speed of sound is equal to the distance traveled divided by the time, $c = s/t$; with the speed of sound c = 340 m/s or 34,000 cm/s, there is a delay t of $\frac{1}{1600}$ sec.

As the sound source moves from the side towards the center—in front of the listener—the difference will be gradually reduced until it reaches a zero value, and then as it moves further to the other side the delay will be noticed again.

In modifying this, it must be said that an impression of direction through difference in time only takes place up to 800 cps; this is related to the functioning of our nervous system (refraction time), as H. Kietz has explained in detail (42). It should be noticed further that there is still no certain theory about how we distinguish forward–backward or up–down directions. Only the side-to-side differences are adequately accounted for.

In addition, the sound source at one side of the listener causes a difference in intensity between the two ears, since the more distant ear is in the shadow of the head and there is also a directional functioning of the ear cavities. The head, however, is no impediment for sound impulses whose wavelengths are longer than the diameter of the head. By measurements it has been established that tones lower than about 500 cps give no impression of direction. That is why some stereophonic playback systems have only one loudspeaker for lows and two loudspeakers for highs.

Furthermore, most people have different hearing sensitivity to loudness increase from one ear to the other. A learning process renders it possible to bring the subjective acoustic center into agreement with the spatial center.

Chapter V

THE CONCEPT OF TIME

1. The Microstructure of the Time Function

Our next basis of perception analysis will be an objective evaluation through the physiological equipment of the whole hearing mechanism, and independent of this the motor-physiological use of the time function, as we will see in the next section.

It is astounding how much information is communicated to the human organs of perception in a musical performance. First of all there are the coordinates of the tonal body: pitch and tone color, dynamics (loudness) and the development of these components in time. Implicit in this is information about space relationships in the performance, inasmuch as even a blind man can judge the approximate size and composition of the orchestra and can locate the individual instrument groups, just as he is able to determine approximately the size of the room in which the performance takes place, as discussed above (p. 59).

If it occurred to someone to express all of these parameters of the communicated time function independently in a code system, he would have a sort of secret language with a very rich variety of symbols (Fig. 50). If the sound structure of the music is reduced to the simplest sound units, which could be labeled *information quanta*, one finds an average of 70 units per sec which the peripheral hearing mechanism processes. In comparison to this, a speaker—considered as the information source—normally provides 50 information units/sec (26).*

* In information theory the unit "binary-digit/sec (= bit/sec)" has been introduced.

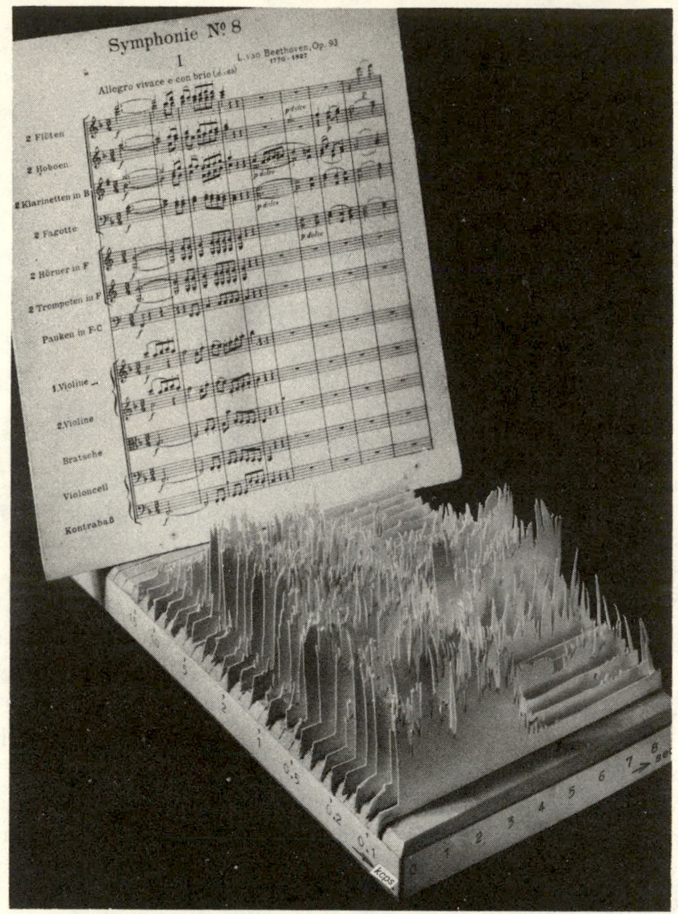

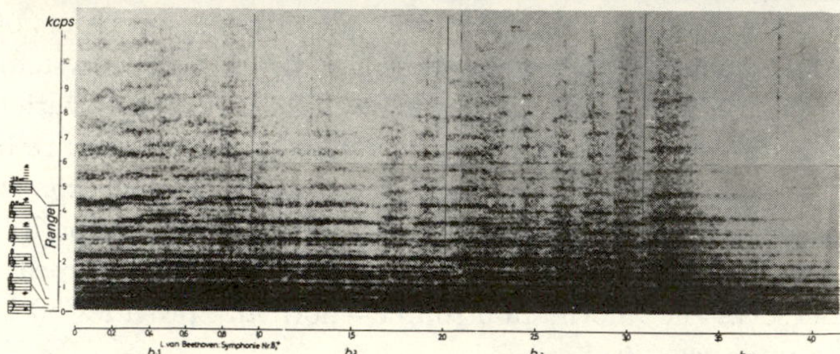

FIG. 50. Sonagram: beginning of Beethoven's Eighth Symphony, displaying the overtone structure up to 12 kcps.

One must remember that the multidimensional acoustical information is transferred to one-dimensional communication channels. This is valid for the ear as well as for electronic recording. It seems miraculous that the tiny point of a needle can simultaneously extract the sounds of a hundred-man orchestra from the hieroglyphics of a recording without mixing the various parts up in total confusion.

Further, we should be aware that in the perception of music we are less sensitive to the time processing of musical events than to the spatial content. Nevertheless, music is only communicated to the organs of perception in the form of time functions (see also p. 78).

The impression of the size and qualities of a room also is produced by means of a time function, for at the position of the listener in the room delayed sound reflections from the walls are added to the primary sound impulses from the sound source (see p. 58), as can be seen in Figure 43. The series of reflections are not perceived individually by the ear if the time interval between them is not greater than 30–50 ms. Since in a large hall the time delay of reflected sounds is greater than in a small room, the first reflected impulses occur relatively later, and in this way the ear distinguishes between the large hall and the tiled bathroom despite nearly similar reverberation conditions.

Impression of location as a function of time is not, however, confined to the ear, for the eye processes visual impression as a time series. The process of looking at a picture is a continuous wandering of the eyes over the surface of the picture, in which the light impression upon the raster-like retina elements continually changes. It is impossible to form a visual impression by means of a fixed stare—even ignoring the fact that in such a case immediate fatigue sets in.

For tactile perception Békésy (A) has shown that the process of impressing a needle on the skin evokes a potential difference, but not if the skin is permanently depressed. These experiments on the forearm served as models for the analogous behaviour of the thin basilar membrane of the inner ear.

If we observe further the processing of time functions received by the ear as they pass through the inner ear to the central part of the brain, one can detect in this system a combined neural analysis according to time and place. The sound excitation is analyzed by the basilar membrane according to location on a frequency scale, then the pitch identification occurs more accurately in a correlation or coincidence process in the nervous system, while in higher regions, by a process of impulse reduction, the chronological impulse figuration appears to be transformed into a local ordering in the nerves (53). The perceptions of the other sensory organs probably take place in a similar way.

It is important to keep in mind that in acoustical perception space and time, with their total of four dimensions, play complementary rôles. Here we have something analogous to what Wells observed in the theory of relativity: "There is no difference whatsoever between time and space, except in the sense that our conscious moves along the time axis." The perception of the stationary room excited by an acoustical event occurs in the microstructure of the statistical reflections, which reach the conscious through a process of integration. Beyond the limit of 50 ms, impulses will be perceived separately (p. 54).

This also explains why sounds above 20 cps, which are series of vibrations not further apart than 50 ms, are not recognized as time-phenomena, but as their own quality of "tone perception." Fluctuations in their periodicity are emotion-stimulating, as was explained in detail on p. 54, and bring into being the disposition for a subjective time sense.

2. The Macrochronic Concept

The macrochronic process, which includes the series of tones, is also not perceived primarily as the course of a time plan, as long as the requisite rhythmic formation is given, binding note to note with regard for the reverberation time—in short, maintaining the flow of the whole. The "whole" is a "gestalt" in the sense of Gestalt psychology; it is perceived instantaneously, independent of time, just as thought, according to the statements of the French philosopher Charles Henry (25), occurs about 100 million times faster than the speed of light. A significant circumstance in music is, however, precisely the communication of a thought which takes on a "gestalt," and the pleasure of musical perception consists partly of inspiration, originating with the composer and communicated by the interpreter. Inasmuch as thought is by far the fastest event in nature, one can make oneself independent of one's space-time environment. "Thoughts are perhaps the strongest realities, the strongest truths which exist" (25). That is the reason why in the strict sense music permits no feeling for time measurement, and that is also the reason why one can "forget" oneself completely in music.

The performer will find it difficult to grasp that time division in music has only a supplementary supporting function, even though it has been made clear that only rhythm is important, not meter.

This almost certainly derives from the fact that bar lines in music have a dominant influence on the eye; and we learn by counting beats by measures, which after all is absolutely necessary for ensemble playing.

Certainly division into measures has had an influence on composition, for instance when we find a piano accompaniment where the steadily moving eighth notes can so easily be counted out (see musical example, Fig. 103).

This accompaniment, with the measures so clearly divided, also provides a support for the listener, for this "division" announces the coming event, namely the next tone in the melody. If the metrical accompanying figure is not present, then a listener is deprived of this supporting guide, and consequently each following sound is met with greater tension and expectation—if the listener provides the required concentration. The information received is thereby increased and increases even more when in addition the expectation of the tone color of the following sounds is no longer certain, as, for example, when the performance of such a piece of music is not restricted to the piano. Such a development was initiated by Anton von Webern (1883–1945; musical example, Fig. 51) and has been recently led to extremes by John Cage. The developments in electronic music are closely related to the above. In compositions like Stockhausen's *Gesang der Jünglinge* or Varèse's *Poème électronique* it is surprising to what extent the duple metrical order remains unnoticed through avoidance of accent, which is typical of electronic music.

In the classical aesthetics of music an increase in the sequence of events per unit of time was felt only as an increase in tempo. The experienced "event" is thus to be understood as its psychically released energy multiplied by time. This means that audible events of a higher spectral density and a succession of events of greater variety, e.g., large interval leaps or variations in tone color, cause us to experience a quickening in the passage of time, that is, a faster tempo. Pauses between notes, however, are also a significant element of the expectancy—this holds true for serial music as well—and we have to calculate them in terms of information theory. A new evaluation must be made of the psychical charges of acoustic -aesthetic structures, and this will finally lead to a new aesthetics of art.

Since the psychical charge is coupled directly to the physical, the question arises whether man is capable of absorbing information up to the limit of his capacity—without neutralization

FIG. 51. Webern: *Concerto*, Op. 24, 2nd movement (Universal Edition).

through redundancy, which in art means through *formal* working out of structures. At the same time this is a law of nature which even the composer of serial music obeys when he makes use of established forms such as repetition, variation, passacaglia and so forth.

As an opposing example, however, one could say that when listening to music the concept "time" appears when the psyche is too little occupied, when too little happens. The German words for "boring" and "entertaining" (*langweilig*, *kurzweilig*) contain this element of time.

Each of us has had the experience during a time of inner excitement of time going by more quickly. When in music we are especially impressed by strong dynamics, changes in color and rhythms, dissonant effects and so forth, time no longer elapses according to the metronome, but rather according to the faster beating of the pulse. Accordingly, rhythms can be pleasant or unpleasant, as Rameau noted as early as 1722 in his *Traité de l'Harmonie*. This French musicologist had possibly the greatest influence on the further development of the tonal functions of harmony.

3. The Measure of Musical Time

(a) PSYCHICAL TIME

Referring again to the fruitful work of F. Brock (13), we can compare the absolute objective units of time with the subjective time (*Eigenzeit*) units of the acting and reacting subject. Instead of employing sidereal dimensions as a point of departure, as in the conventional measuring systems of classical physics, Brock says that for subjective occurrences inside the subjective "stage" (*Eigenweltbühne*) we should refer to the achievements of the subject, namely the marks and effects of a specific protoplasm, which finds itself in continually changing form throughout generations.

Saint Augustine felt the discrepancy between the astronomical unit of time from which the division of the clock was derived and the human pulse, which provides a different feeling of time, according to its individual rhythm. In his *Confessions* he says: "In te, anima mea, tempora metior"—in you, my soul, I measure time. This reminds one of the humorous scene in Hebbel's *Herodes und Mariamne*, where at the court of Herod Persians appear who serve as clocks by counting their pulses.

Here we come upon secrets which are suggested by the upset of the time concept from quite a different direction, that of the theory of relativity. A solution is to be found in the time-measuring physical events which determine the simultaneity of two events, thus including man as a conscious being. The standpoint of the absolute is relinquished.

J. L. Okunewski has presented in a list the pulse count for a number of piano works, in which he measured the pulses of the pianists in relation to the number of breaths per minute. To determine musical effect on the basis of an average number of pulse strokes, as Loulié in 1698 attempted with his chronometer, will not do.

In judging the best time-flow in music we put the following question: How long should a sound phenomenon last in order to be perceived as pleasant; with what speed should tones follow one another? L. W. Stern has called the "time required for an impression to assume its full psychical growth and effect" the *psychical presence time* (*psychische Präsenzzeit*), and has ascribed to it a time value of 0.50 to 0.52 sec (96). Further observations were then made by D. Katz; see the bibliography (40).

However, even this value is not of great significance, for music never is concerned with absolute time and a "gestalt" is formed over a longer rhythmic unit. We must therefore attempt to find a relative evaluation of different note values with respect to one another, according to the human measure.

As early as 1865, the physicist and philosopher Ernst Mach indicated that in the case of two sustained tones played after one another and having only slight difference in duration, the ear cannot determine the longer one from the shorter. When, for example, a whole note is contrasted with a whole note extended by $\frac{1}{64}$, it will scarcely be possible to tell the difference, but one certainly will notice the difference if a $\frac{1}{32}$ note is increased by a $\frac{1}{64}$ note. The perception of the difference depends therefore on the relative lengthening of the time values; perception does not, however, follow a logarithmic law, as Fechner maintained. C. Stumpf objected to this as early as 1883 (99). However, if one takes tones of very short duration, then an arithmetic scale is applicable, as P. Boulez discovered with his montage of sound parts on magnetic tape (12). Loudness is a further variable in this perception. If two tones of equal duration but of unequal loudness are given, the louder tone is perceived as longer, as Stumpf also established. Thus the simple time perception law is of no real importance for musical perception.

In considering the practical necessity of bar lines for the player, which were not employed in earlier times, one must not forget that through metrical organization of the time, the grasping of tone series as a "gestalt" is made easier as a simultaneous act. Still, we cannot entirely agree with Henri Bergson when he thinks of simultaneity and succession as opposites, since through the metrical organization of duration the character of succession is lost and that of simultaneity is gained.

The absolutely metrical treatment of a composition excludes time-gestalts as parametrical components for the perceived whole. As opposed to this, according to an idea of Boris Blacher, varied meter would not suggest time feeling but would contribute to the form (*Gestalt*) of the musical work of art. When Stravinsky postulates for music the absolutely metrical treatment of tone series, he is—not intentionally—limiting the formal means.

(b) FLUCTUATIONS OF THE METRIC SYSTEM.

Let us now observe fluctuations of the metric system. Of course we mean only those fluctuations which are connected with an emotion or, in a deeper sense, biological ones, and not those fluctuations which result, for example, from imperfect playing.

The present author has undertaken the investigation of metrical fluctuations in recordings of different musical works by means of a metronome. Deutsche Grammophon Ges. No. 29108 was employed as a control (Hindemith's Third Quartet, Op. 22, in which the composer himself played the viola part). The movements do not have exact metronome markings, but rather fixed metronome areas, e.g., first movement (Fugato) ♩ = 58–69. A fixed metronome beat is only momentarily synchronized in the course of the music, at most for one measure, since one cannot hold to a fixed meter even for short stretches. The basic measure simply runs from slower to faster, or conversely. It should be mentioned here that in the actual performance of the first movement of the above quartet Hindemith struck up a tempo of 69–100 and during the sections specially marked "livelier" he went up to 138, giving meter variations in a ratio of 1:2. It is as if a musician guarded against being synchronized with a strictly periodic speed, but rather played around the speed within certain bounds, employing such fluctuations for personal expression.

What points of contact do we have with constantly increasing and decreasing tempo? First of all, these variations proceed hand in hand with the dynamics, but they also depend upon the sensual effect of sound, upon the concentration of the sound material, which is the same as the concentration of the spectral lines, and on the difficulty of the manual execution of the piece.

(c) RHYTHM

> *The musical work of art appears only through a movement in time. The specific form of this movement is called rhythm, insofar as it is regulated and structured according to a law.*——Sievers, 1901.

It is not my intention here to discuss once more the problems of defining rhythm, but simply to proceed from a point of view which will help us pursue our goals in the larger scheme of this book. Let us begin with the definition by the philosopher Klages: rhythm is the recurrence of similar material in similar time segments. In our efforts to grasp the measuring technique involved in these concepts, we do not arrive at a sharp definition of meter, but, just as in pitch definition, we arrive at the borderline of uncertainty, which we could fix as a band width of fluctuation. When we apply this to a series of tones belonging together in a piece of music, we see that we cannot determine in advance from their meter the exact beginning and ending in time of the individual tones, for that is a function of other factors, especially the pitch movement, as we shall see below. We can now set up a grid macroscopically—as on the microscopic level (see Fig. 78)—in which the tone is indicated through the extension of an element of the grid, namely through the time fluctuation along the abscissa and the frequency fluctuation along the ordinate.

Still, there are examples in musical literature where strict meter, as the highest expression of order, is set off in some places against rhythm, as the expression of a freer feeling for life: e.g., Beethoven, beginning of the Allegretto of the Eighth Symphony. In this example there is, of course, the expression of the composer's joy over the newly invented metronome of Johann Nepomuk Maelzel (1772–1838).

At the root of rhythm lies a basic perception of motional energy, which in its simplest form pulses in us noticeably as the feeling of walking. Accordingly, rhythm would be a transformation of kinetic sensation into the physical motional sensation of the walking-measure (47). The rhythmic impulses perceived by the ear are led over the nerves for the most part to the bulbar system of the medulla oblongata, while the emotional reaction to the course of the music occurs in the diencephalon and the mental processing of this for comprehension occurs at the highest level, in the cerebral cortex (36).

The result of gathering together all these factors is the increased tensional power (or conversely, relaxation) stimulated by imagination and fantasy. Increase in voluntary tension occurs basically as a muscle tension throughout the player or singer's body, especially in the muscles of the larynx and the muscular complex of respiration with its tributaries to the extremities. Musically, this is translated into an increase in loudness, combined with sharpness (emphasis in the higher overtones). Further, it results in larger intervallic leaps which rise out of a monotone, stimulated by muscular power (cf. the difficulty—like that of an athletic feat—in large intervallic leaps in singing). Thus a feeling of movement is often transferred to the listener. In this survey we have seen that the complex changes in a piece of music under the ordering function of rhythm find their origin in a unified, if equally complex, physiological function.

Having initially attempted to relate rhythm to the world of biological functions, we recognize in our own area of practical music that *breath* is the liaison between the excitement of feeling and the physiological effects.* The trained singer especially will feel this, since he must form the tone on the breath as a modulating process, and his success—apart from the mastering of the basic technique—is qualitatively dependent upon requirements in the area of the soul. The flow of the breath-rhythm, as registered in curves by a physician, is in a certain way a basic measure for the individual formation and interpretation of a work of art. Of special importance again is the "human factor," the time constant which has already played such a great rôle in apperception. It is conceivable that some animals with an entirely different biological "factor" are not capable of following the rhythm of our music.

The functional interdependence of musical elements—melody, harmony, timbre, dynamics, rhythm—does not prevent a composer from making use of the artistic means of occasionally pitting these elements against one another, for example, increasing tension,

* This does not mean that we support the thesis which has been put forward since Democritus that breath is the same as the soul. E. Schrödinger has pointed out this error (88).

through increased loudness, but with decreasing tempo (agogics). Continual change of meter is as important a means of dissolving strict periodicity as, for example, the modification of sound formation through modulatory processes with primary dependence on the onset transients. Still, both phenomena are closely bound together, since increasing tempo means shorter duration of the individual tones, and this shortening means an increase in the noise content and also in volume.

* * *

We have now come to the significant observation that one criterion for the musical impression of a tone series is the uncertainty of frequency and time division. We can bring both together on a graph of uncertainty relation. All other musical elements are to be ordered on a lower level. On the other hand, their permissible fluctuation width from the point of view of perception can be determined experimentally. According to aesthetic laws of art, certain limits may not be exceeded. For the aesthetic impression of a painting, the painter intends that the contours of forms, identified through color areas, should not be absolutely sharp, and that no color should be used in the absolute purity of its spectral line. These requirements are indeed fulfilled in the case of a highly worthy work of art. There is no raster with elements of mathematical accuracy and no monochromatic color value in the execution of painting. The movement of the continual transients is created by the eye in its continual movement over the surface of the picture. A fixed staring at a picture to be defined as quasi stationary prohibits the forming of an aesthetic-artistic impression, the arising of a "gestalt."

We can make similar observations on looking at architectural drawings. In the total representation of houses and gardens—in projects which should be illuminated from the aesthetic side—one avoids the straight line and uses instead a sort of shaded or broken stroke. Franz Liszt once said: "Why subject art to systems of straight lines?"

Chapter VI

THE EVALUATION OF SOUND THROUGH THE HEARING MECHANISM

The music of the future is directed toward a third ear.——F. Nietzsche.

1. The Construction of the Ear

After having dealt with the formation of sound in terms of physical and psychological laws, we arrive now at a more detailed study of how the ear evaluates the objectively formed sounds. These connections, which have been studied in the sciences of psychophysics and psychophysiology, reveal curious anomalies in the occurrence of sound in music, which are not usually predicted by composers, and which can endanger the effect of the reproduction of their compositions.

In order to avoid overextending the discussion of this large area, we will omit a detailed description of the mechanism of the ear, and will consider the outer, middle and inner ear simply as a transformer of the vibrations in the air, carrying them by mechanical and fluid transportation to the excitation-emitting chemical-electrical receptor (Fig. 52).

The quick pressure fluctuations in the air, which one perceives as sound, produce vibrations on the ear drum. These are then transferred across the chain of bones, hammer-anvil-stirrup, to a further membrane, the oval window, which is at the end of the fluid-filled inner ear. Imbedded in this is the basilar membrane, 1.3 in long, and coiled $2\frac{1}{2}$ times in a spiral, rather like a snail's shell.

If one imagines this coiled membrane unrolled, one sees that the pressure fluctuations transferred to the oval window—through the fluid, around the end of the basilar membrane and back again across the second, the round window—are equalized. A fluid wave wanders along the membrane and causes it to bulge accordingly. Opposite the basilar membrane are many rows of sensory cells which are mechanically excited at those places where the basilar membrane is bulged out. Nerve ends are attached to all the hairlike sensory cells of the organ of Corti, which conduct the local excitations to the brain as electrical impulses. This system is stiffened by many supporting cells; the hairs are covered with a membrane.

Helmholtz had thought the frequency fixing mechanics of the inner ear more simple: for him the basilar membrane consisted of rows of taut fibres, tuned to different pitches like piano strings. When a tone entered the ear, the string of corresponding frequency would resonate symphathetically and the resulting vibration on the

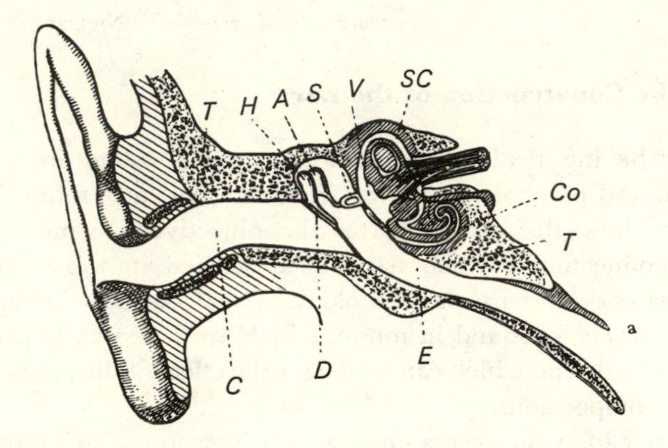

FIG. 52. The human ear (from W. H. Westphal, *Physik*). C = ear canal; D = ear drum; E = Eustachian tube; T = temporal bone; H = hammer; A = anvil; S = stirrup; V = vestibule; SC = semicircular canals; Co = cochlea.

basilar membrane would excite the nerve ends. He thought of the membrane as being formed with the higher tones near to and the deeper tones away from the oval window.

This superseded resonance theory of Helmholtz is compatible with ours only insofar as there is a localization of frequencies on the basilar membrane (*Einortstheorie*), with the sensitivity moving towards low frequencies approaching the end of the spiral. The membrane becomes gradually broader towards this point.

We know today that this membrane is not taut, but is rather more like a stiff plate, and therefore actually is not a "membrane." As a result of this the bulges caused by a sine wave cannot result in a sharp point, exciting only one nerve fiber of one frequency, but always produce slight dents. Accordingly, the dent affects an area from two to three octaves. The copying of the frequencies on the basilar membrane is therefore quite uncertain and consequently the question arises through which process the accurate selectivity is achieved which permits the ear to distinguish two tones that are very close to one another; that is to say, how does the ear determine pitch with an accuracy of 0.3 per cent in a middle frequency range? This will be, e.g., in the case of $c^3 = 1000$ cps, an accuracy of 3 cps or $\frac{1}{40}$ of a whole tone. The frequency selection must somehow take place by a sort of switching mechanism in the nervous system. It is worth noting that many nerves may be attached to one sensory cell, and, conversely, many sensory cells may be attached to one nerve. This could be a switching system similar to an automatic telephone system, as may be seen in Figure 53 (59). The effect is called *lateral inhibition.*

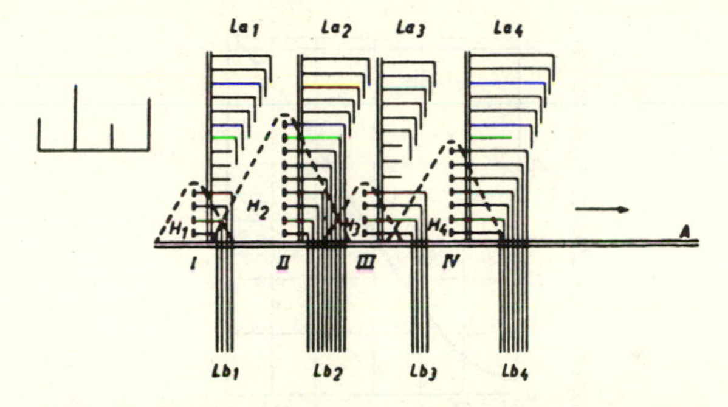

FIG. 53. Quantification scheme of stimulation transmission. $La_1 \ldots La_4$ = unexcited lines; $Lb_1 \ldots Lb_4$ = excited lines; A = basilar membrane path with four areas of stimulation, I...IV; $H_1 \ldots H_4$ = excited hair cell groups of the frequency areas $f_1 \ldots f_4$. Left: spectrum of the selected sound.

This sort of analysis is also not completely compatible with the classical Fourier analysis, since we are not concerned here with standing waves, but with traveling waves; which follow their own physical theory. Figure 54 shows the bulging according to frequency along the unrolled membrane.

The localization of frequencies on the basilar membrane can be seen in Figure 55, derived from anatomical measurements by J. C. Steinberg (118). This diagram shows that the frequency scale

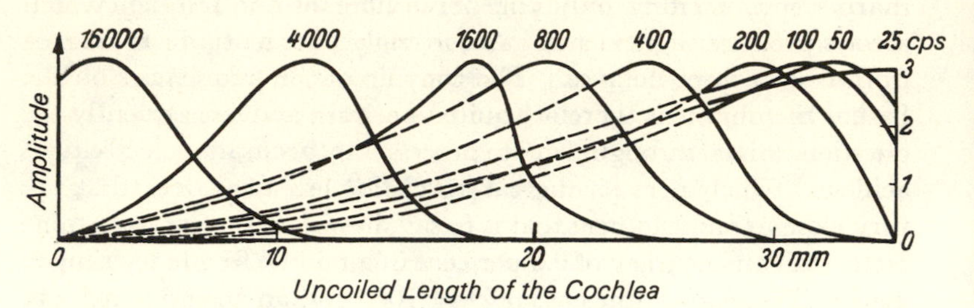

FIG. 54. The bulging of the basilar membrane as a function of frequency.

below 500 cps is nearly linear and above 500 cps proceeds logarithmically (the abscissa is the logarithm of the frequency). The achievable selectivity of pitch perception will be discussed below (p. 95 f.). It is bound together with other physical laws.

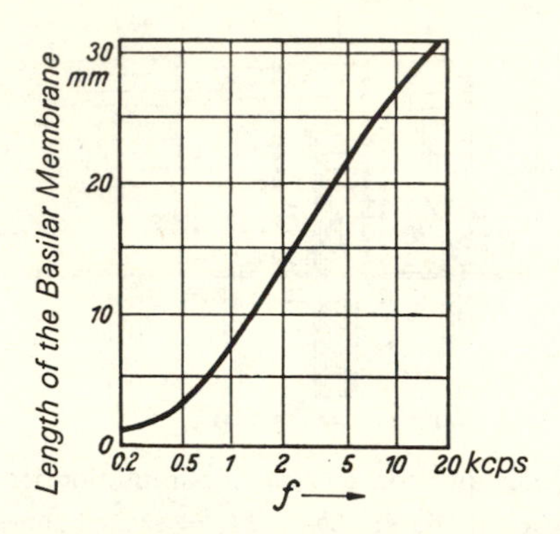

FIG. 55. Localization of frequencies on the basilar membrane.

The goal of this short and simple presentation was to show how many steps of transformation an acoustical process must go through before the receiver is made aware of it. This has to do with physiological apparatus of acoustical, mechanical, hydraulic and electrical types, which obey physical laws and which have technical limits.

In Figure 56 the attempt has been made to represent the complex transformations of the whole aural transmission process; the functions have been given physically, biologically and in the language of information theory. Here one can see how an acoustical event in the environment causes in the sensory organs a stimulation which is coded into other signals, in order to travel over a network of nerves to that portion of the brain in which perception takes place. The latter then makes cognition possible—a phenomenon about which we still know next to nothing today.

That which in musicology up to now has been treated as a psychological phenomenon of music listening is often only the technical characteristic of one of these parts of the whole hearing mechanism and occurs especially when the system is overloaded and when the boundaries of the "operational area" of this sensory organ are violated—in musical terms, when the psychologically acceptable limits of an average tessitura and dynamic range, as well as durational limit, are exceeded.

It would lead us too far astray to list here all the influences caused by each of the single apparatus. We shall discuss only that part which in music produces the strongest distortion in the transformation: the basilar membrane.

Through some simple experiments it can be shown that there is no absolute measure for pitch and dynamics for the ear. If a tuning fork is excited at some distance from a listener and is then brought directly before his ear, he will notice that the pitch at the ear appears lower than from a distance. One experiences this more easily when one sings the perceived pitch with the tuning fork taken as a standard. The dependence of pitch on loudness has been investigated quantitatively by S. S. Stevens, and was represented in a diagram (Fig. 57) (97). The test person was presented with two alternating sounds of identical frequency but with different loudness levels; then one sound was retuned until both tones seemed to be of identical pitch. The percentage of retuning for certain initial test pitches between 150 and 12,000 cps is charted along the ordinate as a function of the loudness. It is amazing how much the pitch of low tones is reduced with increasing loudness and that of high tones is increased.

The extent to which these curves are generally valid cannot yet be determined today. One must assume that there is considerable individual variation.

In explaining the phenomenon presented in Figure 57, one must imagine that the effect of louder tones in a low pitch area causes greater bulging of the basilar membrane, which then becomes stiffer or harder so that the velocity of the wave increases and the

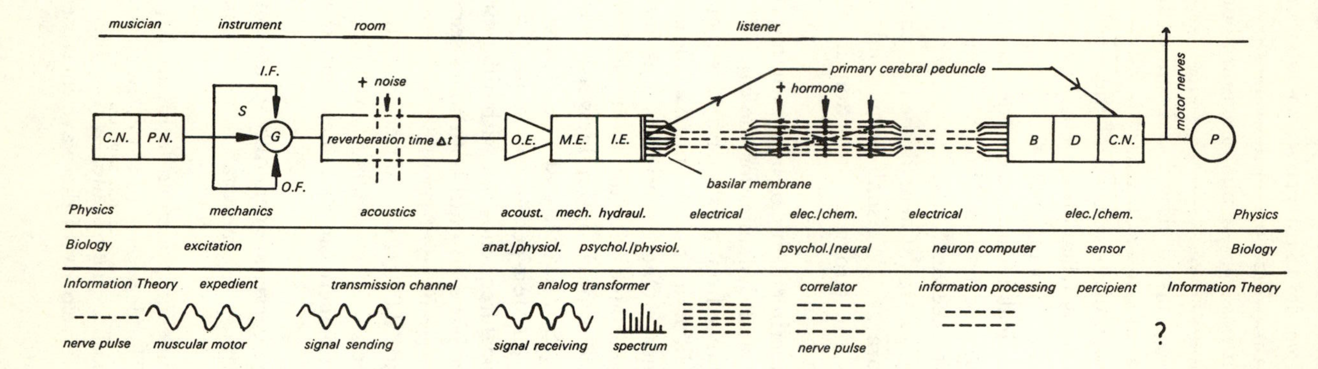

FIG. 56. Diagram of the transmission of acoustic information to the perception center in terms of physics, biology and information theory.

C.N. = central nervous system; P.N. = peripheral nervous system; G = generator (instrumentalists); I.F. = inner feedback; O.F. = outer feedback; S = stimulation; O.E. = outer ear; M.E. = middle ear; I.E. = inner ear; B = bulbar system; D = diencephalon; P = perception.

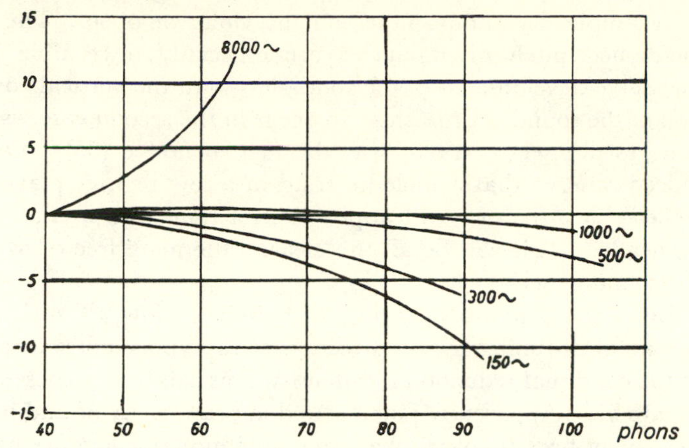

FIG. 57. Contours showing how pitch changes with intensity for sine waves
(according to S. S. Stevens).

wavelength is extended, so that the tones appear to be deeper.* Explanations of pitch migration of loud high tones have not been entirely proven (influence of viscosity or increase of sympathetically moved masses) (81).

Frequent dynamic changes in low and very high tones in a piece of music therefore lead to noticeable distuning, while this will be very small in the middle frequency area. It should also obtain in the case of the single sound, for its spectrum contains a series of partials of decreasing intensity in the proportion of up to 1:100, corresponding to 40 db. Let us assume that such a spectrum with a fundamental of 500 cps has an intensity of 70 db above the threshold of sound. For the sake of simplicity we will assume that six partials up to 3000 cps are being continually reduced in intensity; therefore the formants are ignored. In the case of a reduction of intensity of 40 db to the 6th partial, the distuning from the 1st to the 6th partial should be on the order of 3 per cent. A sound with a fundamental of 150 cps and 70 db intensity would give up to the 20th partial at 3000 cps with 40 db damping a distuning in relation to the fundamental of 10 per cent.

* This involves a variation of Hooke's law, which says that within the elastic limit a strain is proportional to the stress producing it.

The musician does not notice this distuning in practice, when, for example, he intones notes on the violin or oboe. The above mentioned pitch discrepancies are especially discernible in the successive development of the tones, less so in the vertical construction of the sound. Still, they do occur in the second case, as E. M. von Hornbostel (in 1926) was the first to notice (35). One can indeed observe that a melodic triad in a low register played with *crescendo* or *decrescendo* will appear out of tune with a triad played vertically. This will be all the clearer, the more free of overtones the sound generator is.

Even piano sound decaying with time (below c^3) suffers from continual distuning in the direction of the upper notes as a result of the continual reduction in intensity; this can be clearly seen from Figure 57. Apparently piano sound appears pleasant and interesting partly because of its changing intonation, for here we have the continual changes in excitation that the ear requires for aesthetic perception. A similar effect may exist in decay of reverberation, especially in large halls which have a large share of low tone energy in their spectrum. The least distuning occurs in the area of 1000 to 2000 cps, which is relatively independent of loudness. To a great extent one can consider the area of 800 to 3000 cps as relatively sound-stable. Loud tones above 3000 cps at more than 70 db do not occur in music—apart from experiments in electronic music—since they are sharp and unpleasant (cf. the threshold of pain in Figure 77). Distuning in the musical range is rendered less noticeable by vibrato, which causes a pitch uncertainty (p. 107 ff.). A very slight distuning within the spectral structure of sound seems to be evaluated as a positive sensation in the ear.

If several sounds of different frequency occur together, auditory phenomena occur which also can be explained by the behavior of the basilar membrane. The complicated relationships of vibrations, which we will not discuss in detail, lead to distortions which on the one hand can be ascertained to be combination tones, and on the other hand explain the reduction of overtones with increasing frequency as a sort of damping away of the deeper tones by the waves wandering across the membrane. This has been discussed in more detail by Ranke (80).

The dependence of pitch perception on loudness that we have described was first established for constant sounds of sine wave character. If one repeats the experiments, employing short tones with a duration of less than $\frac{1}{4}$ sec, one finds that the tones of all frequencies seem lowered by about $\frac{1}{4}$ to $\frac{1}{2}$ tone (50). This can also occur in practical music when, for instance short tones of changing loudness are accompanied by sounds of other instruments.

This varied behavior in short tone experiments is curious. B. Langenbeck ascribes it to adaptation phenomena. Distuning is also nearly unnoticed when a violin tone without vibrato oscillates between loud and soft (dynamic vibrato). These effects are rendered less acute through another phenomenon. Another effect comes into play to compensate. The perception of the fundamental does not stem alone from the 1st partial of the spectrum, as already mentioned, but usually more from the formation of difference tones from the higher harmonics.

These observations teach us further that sounds with inharmonic partials (bells and triangles, etc.) have no sharp definable fundamental, since the inharmonic series of the partials gives no constant frequency difference. On the other hand, the fundamental appears when harmonic partials are present in addition to inharmonic ones (86).

2. Differentiation Ability for the Pitch Scale

Let us present the following example for the evaluation of intervals (21): On a well-tuned piano strike one after the other *c* and *e*, and afterwards a^3 and $c\sharp^4$; in both cases one perceives the major third as the harmonic completion of the major triad. However, if one thinks of these intervals as parts of a melody, one notices that the melodic difference in the case of the deeper tones in the small octave area is greater than that in the higher octave. This becomes clearer when one avoids a melodic series of tones in which the harmonic consciousness is too strong and selects instead random tones, e.g., B_1–c–e–$g\sharp$ and in the second case $g\sharp^3$–a^3–$c\sharp^4$–f^4.

Perhaps this observation will be made more clear by playing five tones in the middle range of the piano one after another, and then the five highest tones; these last sound much more alike in pitch than those of lower frequencies. We are concerned here with a kind of perception similar to that involved in a color scale, where the gradation of the pure colors is not evenly divided according to the frequencies of the corresponding vibrations.

After questioning several test persons, Feldtkeller and Zwicker

(21) came to the conclusion that in the case of the fifth a^3–e^4 one hears the same melodic pitch difference as in the case of the major third *c–e*.

One can construct a scale of melodic pitch evaluation by always advancing upwards by a certain interval, which seems to be melodically the same as the lowest initial interval. From this we can construct a curve, as shown in Figure 58. The abscissa plots a scale of harmonic partials which proceeds proportionally to the frequency—logarithmically divided; the ordinate plots the evaluation of the melodic pitch, for which the unit of measure "*mel*"

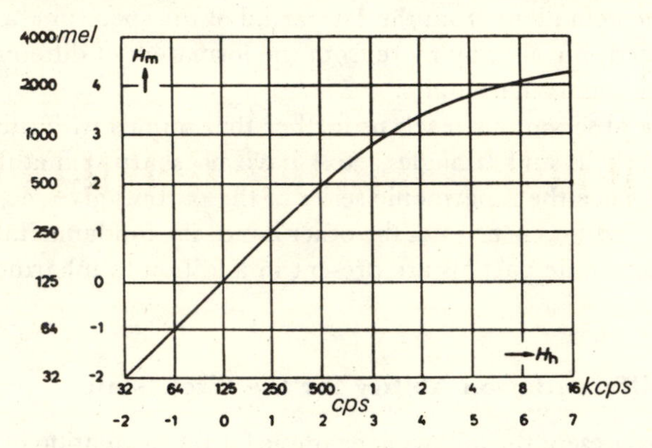

FIG. 58. Relations of melodic pitch H_m and harmonic pitch H_h (the mel scale).

was suggested in 1937 by S. S. Stevens and J. Volkmann.* Up to about c^2 (500 cps) the melodic and the harmonic pitch coincide. Above c^3 one notices a melodic stricture in increasing measure. This provides a further reason to use pitches up to 3500 cps (a^4) at the highest. Seven octaves correspond to the harmonic pitch scale, four octaves correspond to the melodic pitch scale. The octave interval in the highest octave is to be valued at 1:10 rather than the normal 1:2.

The difficulties occur practically when, in writing sequences, a series of sounds is transferred to too high an octave, for then the melodic similarity of the sequence is impaired, while on the other hand polyphonic music presents no difficulty since the harmonic evaluation in all tessituras is based only on the frequency relationship.

* To be sure, with a different reference tone (1000 cps = 1000 mel).

The course of the mel curve provides the reason why the piano tuner tunes with the help of beats only in the range *a* (220 cps) to c^3 (1047 cps) according to the tempered scale, but beyond this

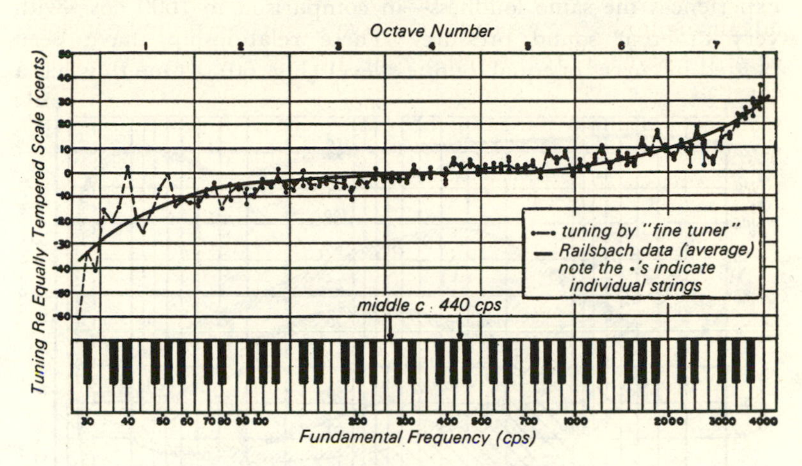

FIG. 59. Measurements on the aural tuning of a spinet piano (according to Martin).

depends on his ability to hear intervals, thereby "stretching" the scale—according to measurements of the Baldwin Piano Company (Fig. 59) (56). The deep tones are strongly dispersed because of the lack of sharp frequency sensitivity (p. 98).

3. The Evaluation of Loudness

Just as in the case of frequency, intensity of sound events is not evaluated uniformly. The limitation exists that the ear does not respond equally to all frequencies.

A tone of 1000 cps is employed as a standard, for which the loudness levels have corresponding sound pressures arranged in a logarithmic scale with reference to an initial value—the threshold value. This proves useful since in this way the great intensity area from the barely audible to the threshold of pain (feeling) can be expressed in the proportion $1:10^{13}$. Thus we have the loudness measure $10 \log I_1/I_2$ with reference to the sound intensity, and $20 \log P_1/P_2$ with reference to the sound pressure. The unit of sound intensity is the decibel (db). If we designate the zero point on the intensity scale as 1, then in logarithmic units of 0 db the sound intensity 10 would be plotted at the 10th scale line, the sound intensity 100 at the 20th, the sound intensity 1000 at the 30th, etc., and finally the intensity 10^{13} on the scale at 130. At other frequencies one experiences the same loudness—in comparison to 1000 cps—with very different sound pressure. These relationships have been plotted in curves of equal loudness level (Fig. 60). One thus has a

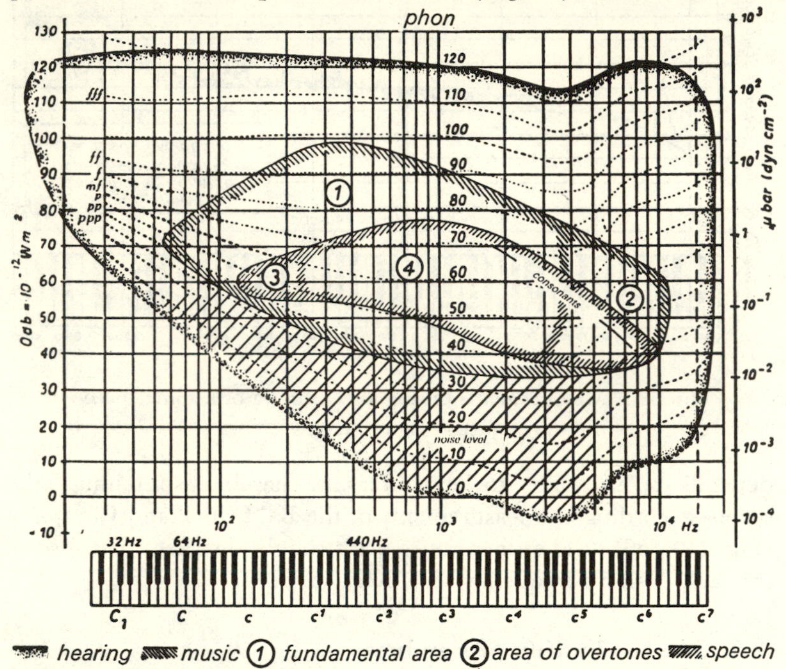

FIG. 60. The area of human hearing. Sound pressure plotted against frequency, with the limits for use in speech and music. (cf. also Figs. 63 and 77).

measure for loudness in all frequencies in the unit *phon*, referring to the scale at 1000 cps, at which db = phon. The lower threshold of hearing is plotted at zero phon, the upper border at 120 phons, where pain sets in.

We can further see from the Fletcher–Munson curves (the dotted lines in Fig. 60) that the ear has a particularly high sensitivity between 2000 and 3000 cps. At lower frequencies the sensitivity of the ear is reduced as the loudness decreases. For a tone of 40 phons (loudness *piano*), for example, one must use at 120 cps ten times the sound pressure as for a tone of 1000 cps. If one sets the volume of a radio lower than it was in the original recording, the lower tones will be preferentially reduced, whereby the tone color will change and the sound complex will become thinner. On the other hand, if one adjusts speech too loud, it "booms," because of overemphasis of the lows. Such distortion can be obviated through an "equalizer."

To be sure, curves of equal loudness level do not give any information about the real perception of loudness, for instance, when a sound is twice as loud as another. Therefore a further unit of subjective loudness, the "sone," has been introduced, with which one differentiates between subjective and objective loudness (Fig. 61).

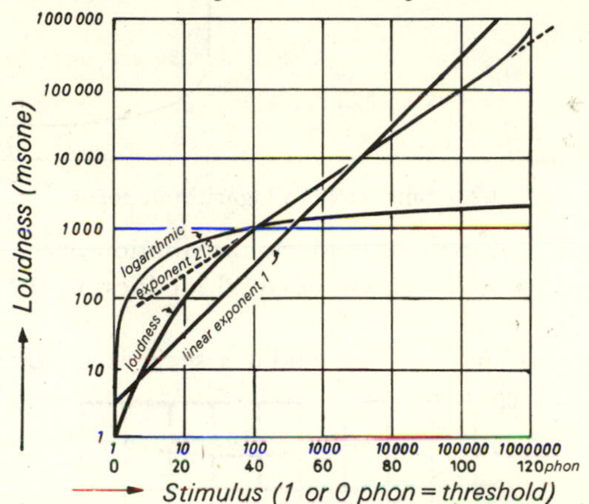

FIG. 61. Loudness sensation plotted against the stimulated loudness level.

Comparison with a logarithmic curve (according to Weber-Fechner), with a linear curve and with an exponential function with exponent $\frac{2}{3}$ (dotted line).

An increase from 30 to 90 sones means three times the loudness.

The comparison curves permit the observation that the perception of subjective loudness is not according to a logarithmic scale, and therefore Fechner's law (1850), according to which perception increases logarithmically with the stimulation, is inapplicable. C. Stumpf had already expressed doubt about this in his *Tonpsychologie*. Subjective loudness perception corresponds most nearly to the curve with the exponent $\frac{2}{3}$. For the levels of use in music, between 40 phons and 100 phons, the best curve would be an exponential curve with the constant exponent 0.67, that is, the subjective loudness level increases as the cube root of the sound intensity relationship (11). In any case subjective loudness follows rather a straight rise than a logarithmic one. In the latter case a sine wave would be as shown in Figure 62, but in actuality it does not happen.

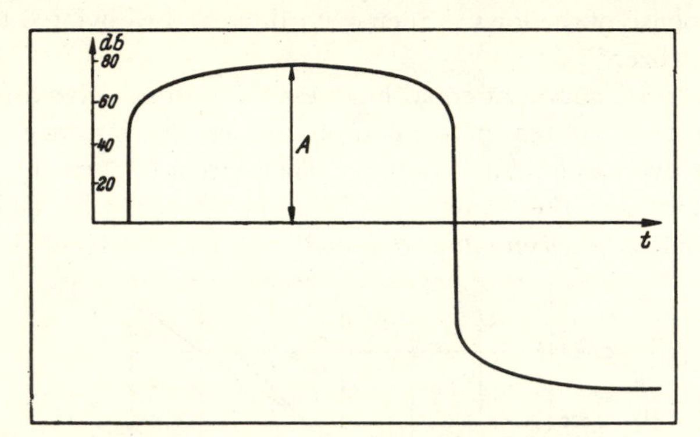

FIG. 62. Sine wave in logarithmic form.

For evaluation of the real loudness distribution over the hearing area one should plot the curves of equal loudness in the sone scale as shown in Figure 63.

Only a limited part of the loudness scale is useful for music.

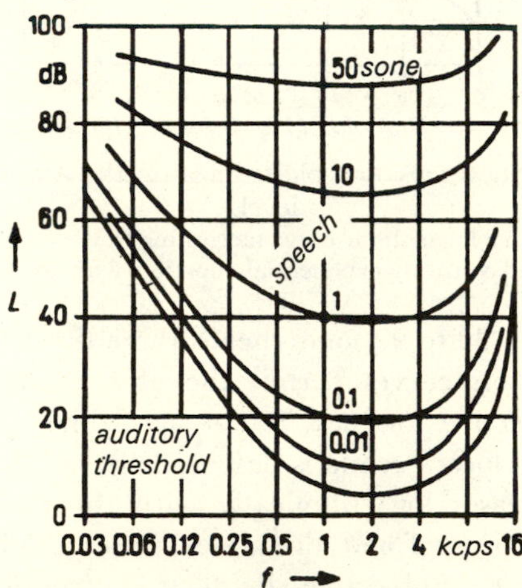

FIG. 63. Hearing area with curves of equal loudness in the sone scale.

Very quiet tones will be masked by omnipresent atmospheric noises in the room. Hence there is the requirement that the signal, in this case the musical sound, be a certain amount louder than the "noise background." The quietest passage in an orchestral concert must have a minimum loudness of 40 phons, or a subjective loudness of 1 sone.

A further limiting of the dynamics is given by the reverberation of the room, which functions as an additional noise background. As Figure 48*a* shows, the signal rises only slightly above the integrating reverberation level.

In the audible field of Figure 60 the most responsive area for musical use has its nucleus in the range of 1000 to 2000 cps, and an extended area of from 800 to 3000 cps is tolerable with respect to purity of sound. The area from 800 to 1000 cps corresponds to the formants of the German vowel "a" (like the "a" in "arm"), which occupies a middle position in the series of vowel colors and is perceived by the ear as optimal. The vocal training of a singer or actor usually proceeds from this favorable vowel, and it has been shown further that the secret of the beautiful sound of the most famous Baroque organs is this special "a" formant which is characteristic of them (111). Also, raising the frequency area by 1000 cps in the reverberation time of concert halls has proved desirable. In addition, the ear is most sensitive to frequency and amplitude modulation in the range of 800 to 1000 cps. This means that when the frequency scale is narrowed more and more, the understandability of language is maintained the longest in the area of the "a"-formant (cf. the measurements of Kryter). For the same reason this area is particularly insensitive to noise masking. A voice leading in a score is also most clear and most prominent in this favorable frequency range. The reproduction will be perceived as most pleasant with a loudness of 70 phons (*mezzoforte*). Thus in the area of all audible tones an optimal region can be found, and this region in Figure 60 is around 1000 cps, at 70 phons..

4. The Transmission of Acoustical Events in the Nerves

The mechanism of transmission of signals in the nervous system has been explained through hypotheses of Wever and Bray (97). In the stimulation taking place in the organ of Corti in the inner ear there is a quantification process, in which for each incoming vibration an electrical impulse is sent off (Fig. 64) and travels through the nervous system to the higher centers of the brain. All impulses have equal strength and duration, regardless of whether we are dealing with a loud or soft or a high or low tone ("all-or-nothing" law).

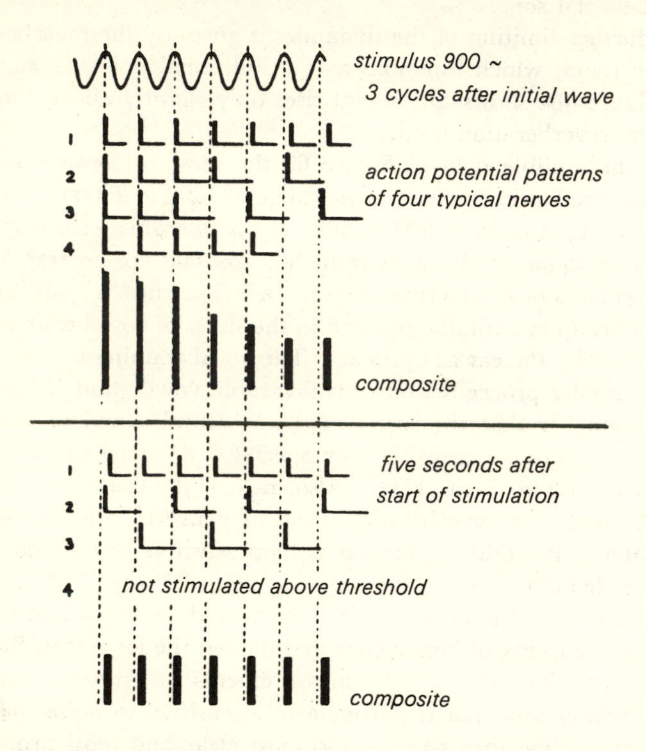

FIG. 64. Stimulation of the auditory nerve fibres according to the "volley" principle (Wever and Bray, 1930).

The solid vertical lines represent the action-potentials of individual fibres. The horizontal lines represent the duration of the functional refractory periods. The heavy vertical lines show the composite effect of the individual action-potentials.

A limit is set by the fact that the single nerve fibres can transmit only up to about 1000 impulses per second, because each fibre must recover from each excitation; this is called refraction time and lasts approximately $\frac{1}{1000}$ of a sec. During this time the system is not excitable for other impulses. For acoustical stimuli of higher frequencies a second parallel nerve fibre operates by taking up every second impulse (Fig. 64). Frequencies from 2000 to 3000 cps are distributed over three fibres. Impulses of more than 3000 to 4000 cps cannot be handled by the help of this synchronization system; the system is overburdened so that for higher frequencies another impulse code becomes operative. Degrees of loudness are differentiated by the excitation of more fibres as the strength of the stimulus increases. This hypothesis is called the volley theory.

Since the formation of the volley theory important new insights have been gained into the function of the central nervous system in response to acoustical stimuli. These hypotheses are based on Electroencephalogram recordings (J).

5. Fatigue and Adaptation in the Hearing System

Although we now have found clear connections between acoustical intensity and subjective perception of loudness, we are not able to determine the subjective loudness for every instant of a piece of music. If we permit a tone of longer duration to be impressed upon the hearing system, the subjective loudness perception for this tone will at first increase, until after about 0.2 sec a gradual decrease in the subjective perception occurs, which results from a gradual fatigue through the constant burdening of the sensory cells and the nerve fibres.

The changes in sensitivity in response to the effect of a constant burden of sound affects the following tones in the influence area of the constant tone for several seconds. One then speaks of "fatigue." Békésy (A) determined the adaptation by supplying to one ear a constant tone and to the other ear a comparison tone, interrupted periodically, by which he measured the decrease in subjective loudness (cf. Fig. 65, top). From this we learn, for example, that one tone following after a loud organ point of 5 sec duration at the same pitch will be heard after $1\frac{1}{2}$ sec with a loudness of 5 db less than the original tone, therefore softer than intended by the composer. Important here is the initial loudness of a constant tone of longer duration. In Fig. 66 one sees that the decline in the subjective loudness is greater for the loudest tone, with the result that the end loudness after 2 minutes is the least for the loudest tone. A very loudly played organ point of constant sound pressure would therefore bring about great revaluation of the loudness of

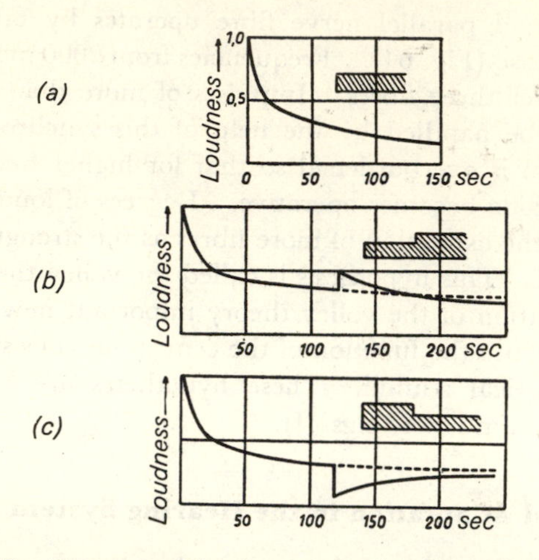

FIG. 65. Decrease of loudness sensation through fatigue when the ear is taxed with a permanent sine wave, measured over 3 min.

Hatched: time course of the test signal. (*a*) Constant sound pressure; (*b*) doubling of the pressure after 2 min; (*c*) halving of the pressure after 2 min.

following tones of the same pitch, as well as neighboring tones in the same area of influence. This decline does not occur in beat tones because the ear recovers after each decay, as Helmholtz observed. In musical reproduction we must always keep this in mind.

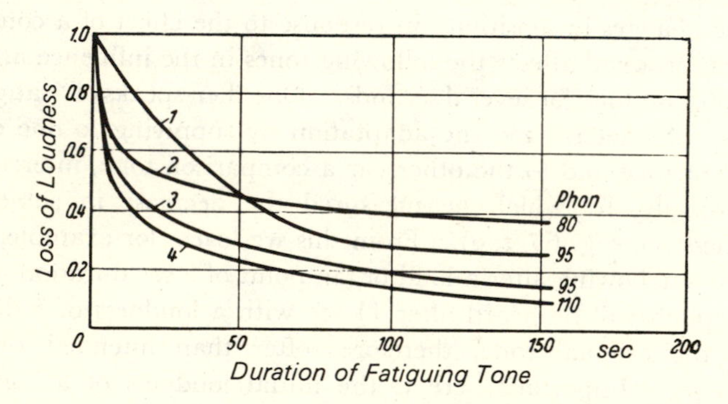

FIG. 66. Fatigue as a function of the intensity of the fatiguing tone (800 cps) for the three loudness levels 80, 95 and 110 phons (curves 2, 3 and 4, respectively).

Curve 1 shows the simultaneous decrease of the differential thresholds of pitch and loudness level or increase in sensitivity to pitch and loudness level differentiation (according to Békésy).

Measurements by Békésy (A) show how with increasing beating rate the fatigue effect declines, and completely disappears for a beat of 15/sec (Fig. 67). Still, this behavior varies with individuals.

It is further interesting to observe that a constant tone, which, for example, swells in 2 minutes to double the sound pressure, finally attains scarcely half of its initial subjective loudness (Fig. 65, center). On the other hand, the reduction to half of the sound pressure results in the same subjective loudness as in the previous experiment (Fig. 65, bottom). One can further assume that an acoustically fatigued ear becomes more sensitive, and might overvalue a sudden change in loudness, as Békésy mentioned in 1929 (A) and Lerche (51) established more recently. These experiments

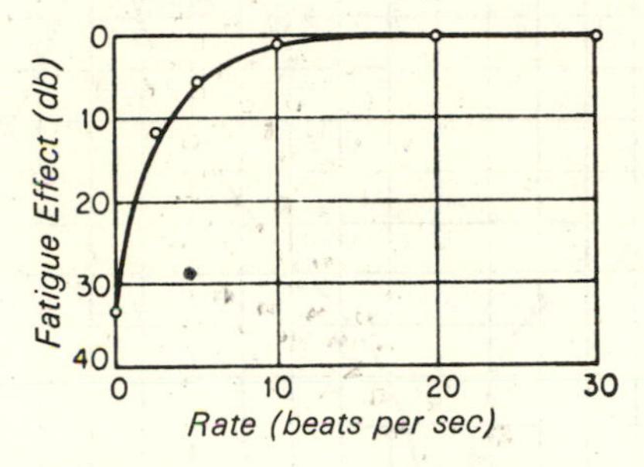

FIG. 67. Effect of beating rate on fatigue. Duration of the fatigue 2 min.

(*Experiments in Hearing*, G. Békésy; © 1960 McGraw-Hill. Used by permission of McGraw-Hill Book Co.)

demonstrate that *a musician has no absolute measure of objective loudness, just as he has no absolute measure of time.* According to the musical context, a *piano* can unconsciously be played *forte*, since the reference level is relative. After a *pp* a *mf* seems like *ff*, and vice versa.

The influence area of the adaptation frequency extends also to the neighboring frequencies in an area of an octave above and below. In the case of a biasing stimulation of the ear by means of an organ point it will happen that a tone a certain interval away will sound out of tune if it lies partly in the influence area of the constant tone, as a result of the localized biasing stimulation of the basilar membrane in conjunction with the groups of sensory cells connected to it. According to measurements (Fig. 68), for a constant tone at 800 cps (approximately g^2), a tone a fifth higher sounds about 7 per cent out of tune, that is a little bit more than a half tone, and a lower tone at 500 cps sounds about 6 per cent out of tune, which is exactly a half tone (A). This can have the effect for an organ point at g^2 of the following octaves c^2 and c^3 sounding a whole tone out of tune. Therefore, in the case of pieces with octaves following organ points, there is not only change in loudness, but also considerable distuning.

It can occur pathologically that one ear loses its sensitivity in limited frequency areas, so that this ear perceives pitches differently

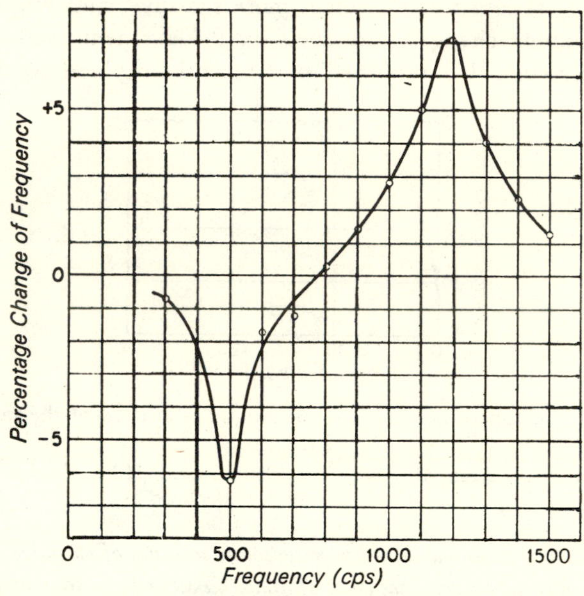

Fig. 68. Effects of fatigue on pitch by distuning in the area of an octave above and below.
Constant tone of 800 cps (according to Békésy).

from the other ear; thus in binaural hearing tones will be heard doubly (diplacusis). C. Stumpf observed as early as 1883 that a minor diplacusis can occur in normal hearing as a result of unequal response in the two ears (99).

The above mentioned fatigue phenomena resulting from the acoustical overstimulation of the hearing system permit one to recognize the significance of a general rest, especially after *fortissimo* passages, as a time span necessary for the restoration of the normal hearing response. One must remember that during a part or the whole of the rest the sound energy decays according to the time constant of the reverberation, and the musical effect of the general rest can lie in this decay.

Figure 66 demonstrates further that as a result of adaptation the relative difference thresholds for intensity and frequency decay in the same way, that is, with increasing adaptation the ear becomes more sensitive to intensity and pitch differentiation (A). This means that in connection with our organ point example slight modulatory changes in the sense of an amplitude and frequency modulation will be more clearly heard for the following tones in a certain influence area on both sides of the adapted tone than without the previous biasing stimulation. One can see from Figure 66 that the highest sensitivity for frequency differentiation is reached only after one minute, while the loudness decline is reached immediately. Because of this sharpened sensitivity, sudden changes in loudness are quantitatively overvalued.

A further effect of an organ point consists in the following: When there is sufficient loudness, higher tones even above an octave will be masked, but not lower tones. A low-lying organ point used as a bass foundation is therefore more significant with respect to masking than is a higher one.

In summarizing, we can say that an organ point which lasts long enough and has sufficient loudness exerts a considerable influence on the tones which follow, with respect to the entire sound structure, because of the combined effect of the reduction in loudness, because of the distuning of neighboring tones, the masking of higher tones on the one hand and the increase in response for the perception of the smallest pitch and loudness fluctuations (inner modulation of sounds) on the other.

6. The Evaluation of Vibrato

In the introduction of this book it was shown that a stationary condition, as a result of its strictly periodic character, is no longer perceivable after a corresponding onset time, and we have now seen what a distorting influence a constant tone, for example, can have on the tonal environment.

Sounds of long duration therefore have disadvantageous results in music and are avoided as far as possible. Where an organ point in a polyphonic composition serves a purpose, e.g., as a carrier of harmony, or as a shift of emphasis of musical texture into the bass, it should be executed—when stylistically possible—with vibrato or trilled; this corresponds to a frequency modulation. Vibrato develops inevitably, especially during the study of string instruments and singing, and its continued use demonstrates that this inner movement of tone has proved musically advantageous.

The curious effect of vibrato has not yet been satisfactorily explained physically. The first thorough investigation was made by C. E. Seashore. At first it seems odd that a continuous change of the frequency should generate a discontinuous spectrum, which under certain circumstances can also be heard as such (Fig. 69).

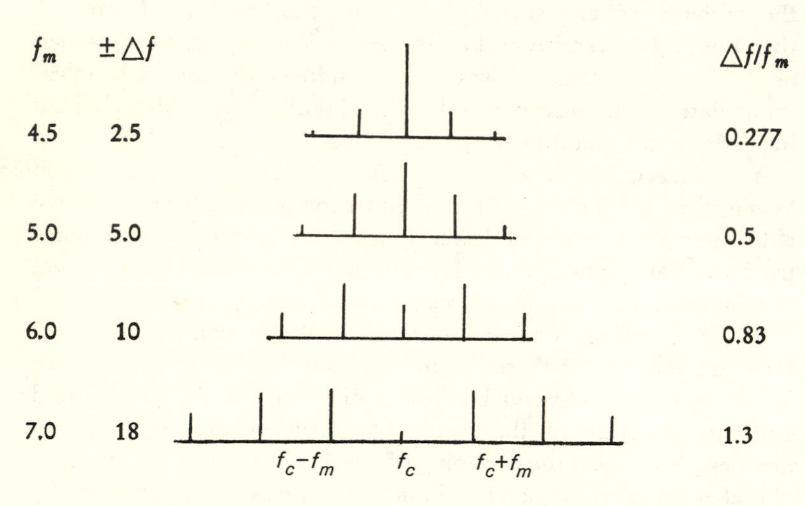

FIG. 69. Spectra of frequency modulation of a simple tone (c^2) at different vibrato frequencies f_m and different frequency ranges $\pm\Delta f$. Computation example: $1.25:4.5 = 0.277\ \Delta f/f_m$.

The fundamental to be modulated has, in the physical effect, the function of a carrier frequency f_c which is modulated with the frequency of the vibrato change f_m. The distance between the spectral lines is f_m. Nothing is changed in this hearing phenomenon by the fact that the spectrum can be accurately derived mathematically from the frequency modulation (N).

For a more detailed investigation we will use a tone generator with a wobble circuit, which permits a periodic change in a certain frequency area for any tone. In the case of an adjusted vibrato of 1–5 changes per second one recognizes the periodicity of pitch change, most clearly in the case of 4 per sec, a value recently confirmed by E. Zwicker in his determination of modulation threshold, according to the observation of Shower and Biddulph as well as Riess (117). In the case of a further increase in the vibrato frequency—beginning with about 6 per sec—one notices only the unequivocal pitch of the initial tone f_1 with intensity fluctuations of the period of the vibrato. In the case of a still higher vibrato frequency—up to 12 per sec—one hears instead of a single tone something like a group of tones. This is already the unpleasant area of the tremolo.

If we extend this observation by modulating a tone f_c of 3000 cps by a frequency f_m of 100 cps, the frequency of 3000 cps as well as that of 100 cps will be audible, providing an experimental proof of the formation of the fundamental as a residual tone.

D. A. Ramsdell has executed systematic investigations in which he asked experienced musicians how they evaluate the adjustment of certain vibrato frequencies and the corresponding various depths of modulation (N). The result is shown in Figure 70. In the places on the curve marked by a small circle only a single pitch was heard, therefore no gliding of the frequency, but a periodic oscillation of loudness. Tones occurring along the vertical dotted lines were considered especially full. Thus at the intersection point of the two curves a tone must be both distinct in pitch and full in sound. Measurements of good professional voices show that the vibrato rate of 7/sec, enclosed in the large circle, corresponds to the value arrived at synthetically. The vibrato of a violin has the same periodic change of 7/sec, but the lesser depth of modulation of 8 cps. This auditory impression of a singer's vibrato has led teachers to the erroneous impression that it is an amplitude modulation. This is indeed present, but only to a negligible degree.

It is interesting that the impression of a continuous change in frequency of seven times per second is perceived as a fixed pitch, that is, when an acoustical event (vibration of the amplitude modulation) recurs continually every 0.14 sec, the value which Békésy (1933) has given for the persistence of tone sensation (A); this would explain the fact that within this time span two consecutive sound events can no longer be perceived separately (N).

Furthermore, it must be noted that in the performance of music

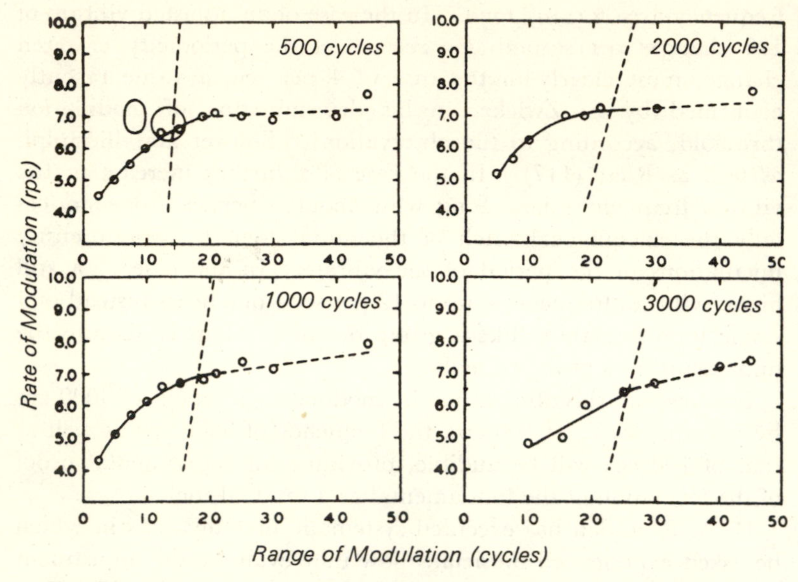

FIG. 70. Determination of the optimal data of vibrato for aesthetic tonal effect.

The critical rates and ranges of frequency modulation producing singleness of pitch (small circles on curves) and maximal richness (vertical dotted lines) (according to Ramsdell). In the plot for 500 cycles, the large circle represents the rate and range of vibrato in the voices of accomplished singers, and the oval shows the rate and range of vibrato produced by expert violinists.

vibrato is not strictly periodic, but has slight fluctuations which mark the beginning and the final part of the sound. This is recognizable in voice registrations according to the Visible Speech method (Fig. 71). Electronic organs have a built-in wobble circuit which gives a strictly periodic vibrato, and which makes the sound rigid inasmuch as a stable condition is formed as in the case of a constant tone of longer duration. Again we see that music cannot suffer any strict periodicity.

FIG. 71. Vibrato of a soprano singer (sonagram).

Although with a vibrato of 6–7/sec one can identify a distinct pitch, the recognition is not so certain and spontaneous as with a pure, unmodulated tone. In this slight uncertainty lies the attractiveness of a development of musical tone. The greater the depth of modulation, the more unclear is the impression of pitch. With the introduction of frequency modulation (vibrato) the perception problem of constant tones of longer duration has not been fully solved. The stimulated frequency area which moves about nearly periodically over the area of the modulation depth always finds itself in the influence area of a frequency group, so that a periodicity problem arises in another form, affecting the perception by fatigue.

7. The Time Constant of the Duration Threshold

We have already established that the ear as a mechanical vibrating system needs a certain time until it is brought into a steady state through the incoming sound waves. This does not prevent the characteristic sound from being recognized as such beforehand. The minimum time for tone recognition is conditioned mainly by the energy content of the sound (21). The minimum recognition time (duration threshold) is 4 ms for sine tones switched on suddenly. If the duration of the tone is still shorter, the hearing system will register simply a click. This leads to the concept of "physiological onset time," which corresponds to the duration of the switching click of 0.25 ms. If the tone does not begin so suddenly, but rather starts somewhat more softly, tone recognition perception in higher ranges—about 2000 cps—functions already at 3 ms (104) (Fig. 72). If the time constant for the recognition (*Tonkennzeit*) and also for the physiological onset were considerably shorter, then all sudden sounds would be intolerably hard, and all attacks in music would have to be extremely carefully executed in order not to cause pain to the ear. The "softer onset"—longer onset transients—occurs, as explained, with a less broad frequency spectrum; there is less noise present, so that the sound actually intoned can be recognized more quickly. As can be seen in Figure 29, within 10 ms the ear can identify the sounds of instruments, even if they are of very short duration, simply by the onset. Toward the low tones, the required recognition time increases, as

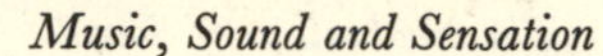

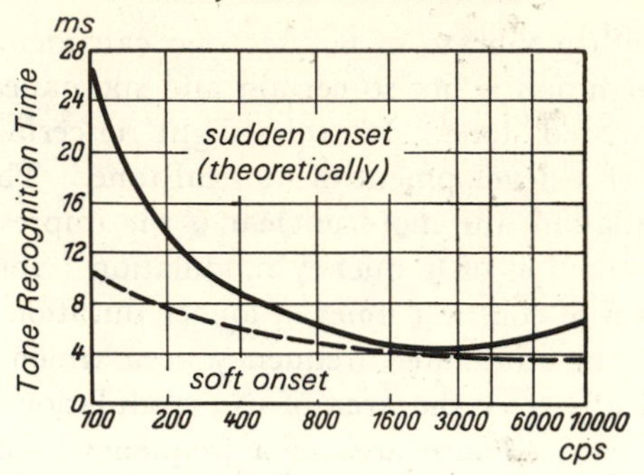

FIG. 72. Delay of tone recognition (sine tones) as a function of the frequency.

seen in the curve, to 10 ms at 100 cps. This is also the active area of male speech pitch.

8. Sound Fusion and the Formation of Color

As already mentioned, it is assumed that an integration across the whole spectrum is attempted in the brain center of hearing; it would be such that in scanning the envelope (dotted line in Fig. 11) the formant resonant peaks would be noticed, in order then to fuse into a unified sound (77). From his investigations on vowel colors, M. Joos estimates the time that is necessary to establish a formant out of a group of harmonics at approximately 30 ms. Since in addition to this there is the time constant of the audible smear of 50 ms (see p. 51), although the color formation has already taken place within the 50 ms, the complete *tone color recognition* time can be given as $\sqrt{50^2 + 30^2}$ = approximately 60 ms (38).

The integration work of the brain becomes greater and lasts longer as more neighboring spectral lines need to be scanned for the integrating impression. In other words, only in the case of tones of longer duration is a complete fusion of the formant peaks contained in the envelope possible (Fig. 11). Short sound processes such as speech or rapid tone series therefore require a slightly damped resonating form of the vibrating system, that is, one which possesses small resonant breadth. This contradicts the requirement mentioned above (see p. 50) that strong damping is in the interest of a rapid onset. It is therefore necessary to make a compromise between fast onset and good fusion. With this knowledge one can develop instruments more in the direction of full sound or more in the direction of sharp onset (percussion instruments).

More accurate reports are available on the perception of vowel color. From Figure 73 one can see how much one may shift the formant frequencies F_1 and F_2 before an original vowel color is so distorted that it is no longer recognized. From this we can derive a two-formant tolerance area of specific sound color recognition.

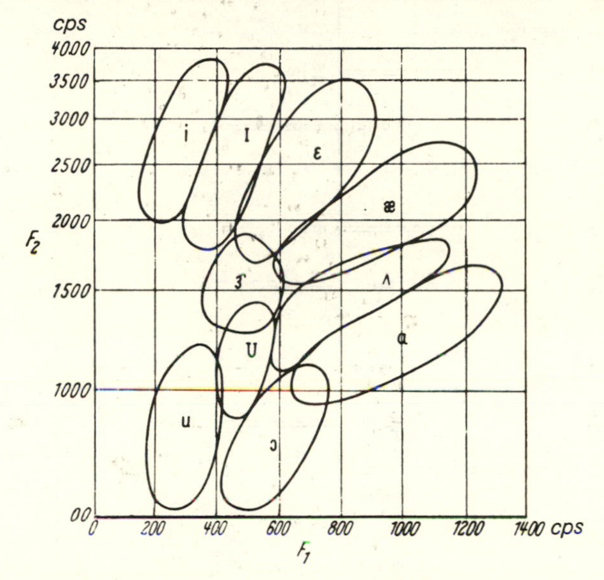

FIG. 73. The vowel characters depending on the formants F_1 and F_2 are recognized within the wide areas delimited on the diagram. (according to Peterson and Barnay).

As mentioned above (p. 19), O. Sala has pointed out that the stability of tone color along a pitch scale remains only when the two chief formants are present.

Further, the amplitudes and frequencies of single partials of a sound spectrum can be changed greatly before a distortion of the tone color is noticed. This can easily be tested today since the equipment for speech and sound synthesis permits the amplitude and frequency of single partials to be altered. Stumpf has already arranged such experiments by eclipsing overtones by interference (98). The insensitivity of the ear to change of amplitude of such overtones is so significant that the erasing of an entire partial,

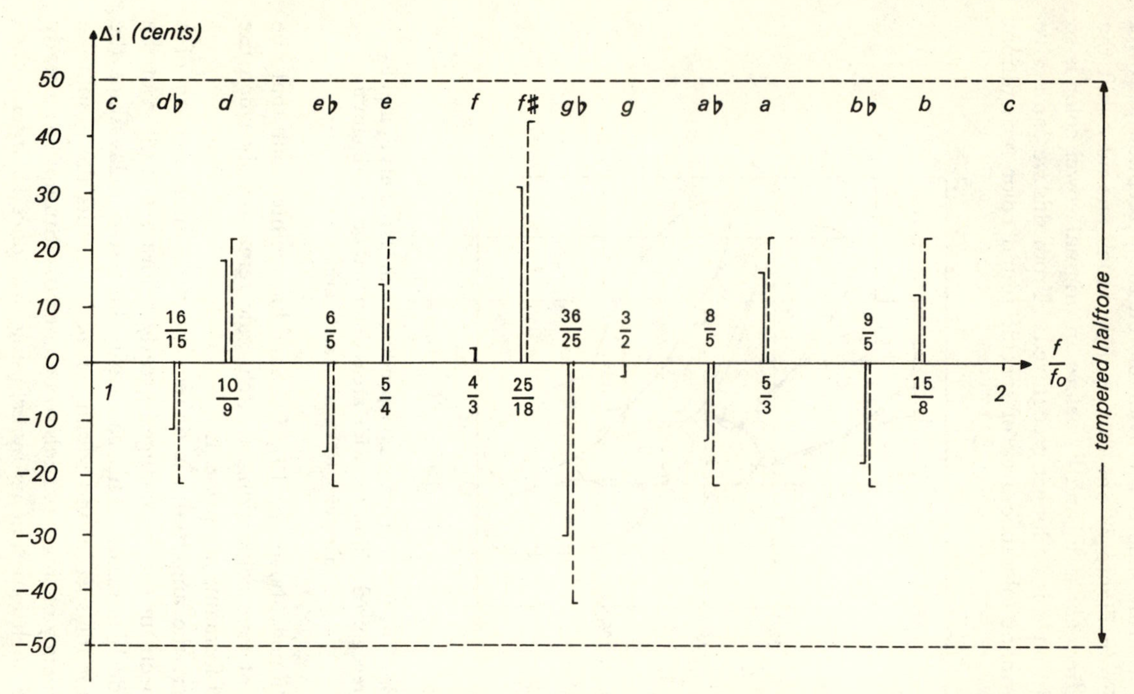

FIG. 74. Pitch deviations of the Pythagorean scale (dotted lines) and the tempered scale (full lines) from the natural scale (in cents) in the intervals within an octave (horizontal line).

as long as it is not a formant, is not noticed by the untrained ear. Ear damage which occurs in noisy factories or as a result of an accident, and which reduces hearing sensitivity from about 4000 cps or below, is often not noticed by the person affected and does not disturb his enjoyment of music.

The ear is also rather insensitive to preemphasis or damping of distinct frequency area. In music reproduction, the middle and high areas of the spectrum can be reduced in amplitude up to 4 db before it is noticeable. In fact, in the low octaves such a "linear distortion" is perceived only at 10 db reduction. It is interesting that these results are independent of the type of music. The ear is much more sensitive when the experiments are carried on in the laboratory with sine tones or with noise; then a trained ear picks up deviations in the frequency curve greater than 1 db. This is a tolerance border that must be matched by good loudspeakers.

The ear is also surprisingly insensitive to frequency deviations of single partials in the spectrum. For example, the Hammond organ is so built that the eight partials employed for the sound are formed from components of the tempered scale rather than from harmonics of the fundamental, which would correspond to the natural scale. The deviation between the two scales is seen in Figure 74.

Little changes of single partial frequencies are therefore possible without destroying the blend of the characteristic sound, since each partial is not a discrete spectral line but has a band width as shown in Figure 5. Tone color beats can occur according to the frequency area, which in the end leads to rough effects.

Finally, for the purposes of electronic music, it has become a principle to shift whole spectra by certain frequency values; this procedure is called frequency transformation and inversion is also employed for coding and decoding speech (110). This is not to be confused with transposition to other scale steps, where the overtones remain whole multiples of the fundamental. This is nothing but a frequency extension or compression of the single spectral lines relative to each other. But in the case of frequency shift the partials maintain their respective frequency differences from one another (Fig. 75*a*; see also Fig. 10). This is achieved by single-sideband modulation with carrier suppression which is explained in Fig. 75*b*. The experiment shows that one can undertake wide shifts without disintegrating the sound complex and without being

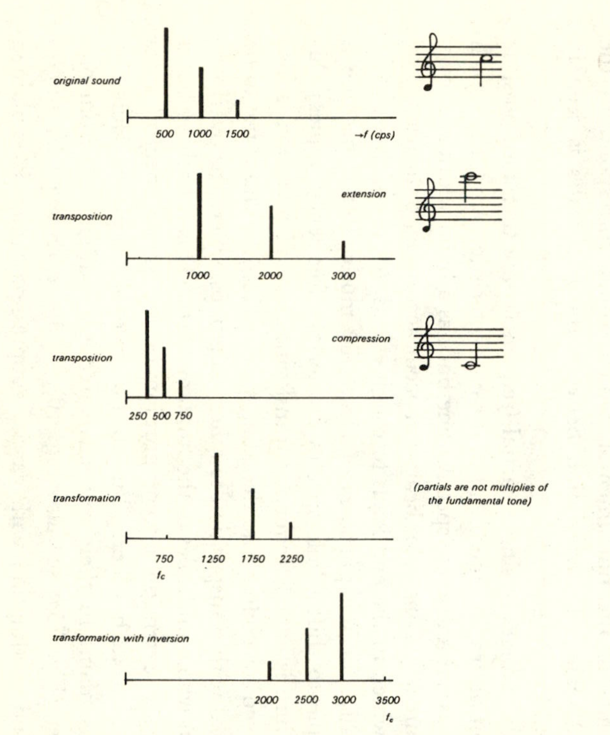

FIG. 75a. Possible alterations of a spectral sound in natural and electronic music.

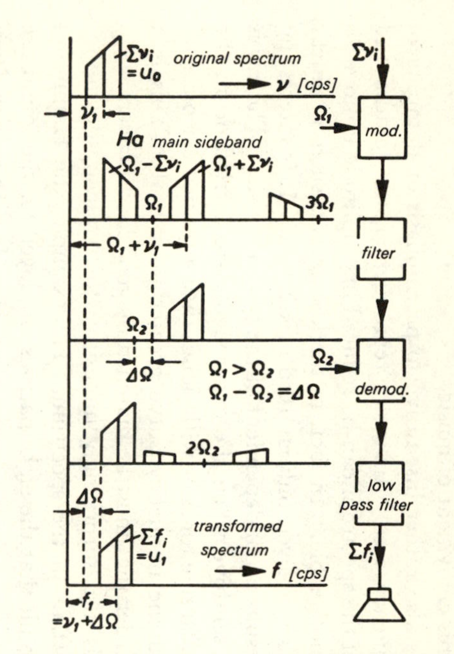

Fig. 75b. Transformation products caused by detuned carrier modulation. ν—spectral frequencies; Ω—carrier frequency (according to L. Heck and F. Bürck).

able to discern single partials. But the tone color in this case is strange, having little or no analogy to the original color. In speech one can execute shifts up to 200 cps without rendering the sounds unrecognizable; of course here the speech rhythm offers a good contextual aid.

Thus it is impressive to observe how great is the power of blending of a complex sound and how difficult it is to split it into its components. In spite of detailed research, especially by C. Stumpf, we are still unable today fully to explain this phenomenon physically or physiologically.

On the other hand, one has to observe how little binding exists between the individual instrumental sounds of an orchestra. In a performance the experienced listener can usually clearly tell most of the instruments from each other. It is quite different with the colors of painters who have infinite mixtures at their disposal. Total blending, when sought after by composers, can only succeed in rare cases, and even then the sound is heard as many-leveled. With constant unison sound there is a certain time constant before the running together of the components can be noticed. Even with electronic sound generation, there has been considerable difficulty in blending independent sounds together. The persistent individuality of instrumental sounds is caused by onset transients and phase relationships.

An attempt has been made in electronic music to develop a tone color scale by creating new ordering principles in the sound structure of a composition, and accordingly by generating sounds synthetically from partials with defined amplitudes and, if desired, with inharmonic frequencies. It is an extraordinarily laborious process to give an individual form to each tone, and the results are disappointing; because of the wide tolerance mentioned above the desired sound variation is not attainable to the desired degree of subtle gradation. For serial music, in any case, one cannot use the same ordering principle for tone color as for pitch, if one is unwilling merely to take a rather arbitrary series of available tone colors.

On the other hand, observation of the transients show us how critical tone color formation is. Sound fluctuation under certain circumstances can cause a noticeable change in color. Transformations, in sound and musical events, ornamentation, sound modulation, etc., can make tone color perception difficult if a certain limit is exceeded, although the relationship of fundamental to overtones remains relatively the same. This can perhaps be accounted for by the time span necessary for the formation of a color impression in the brain.

Naturally infinitely small tone color differences cannot be distinguished, but there is a differentiation threshold which consists of about one half tone for the displacement of the frequency of the significant formant. If we ignore for the moment the fundamental fluctuations, which are synchronous with the overtone fluctuations, we still have to consider the case where the fundamental is invariable, but the formant frequency is varied; this can easily be demonstrated with electronic instruments (84). Tone color disintegrates and one hears single overtones of the fundamental in succession (cf. the mixture stop of organs). A similar case occurs when the intensity of an overtone is constantly increased until it is heard all by itself. Tone color perception is then disturbed (cf. the interference investigations of C. Stumpf).

9. The Connection Between Pitch and Color

We now have acquired conceptions of the hearing process which account for pitch perception, the interpretation of tone color and even the interdependence of these two. The existence of such connections has been suspected for some time; only the physical proof was lacking. Arnold Schönberg in his *Harmonielehre* said (85): "I cannot accept the distinction between tone color and pitch as it is generally stated. I find that tone makes itself noticed through color, one dimension of which is pitch. Tone color is therefore the large area, of which pitch is one division."

H. Helmholtz and C. Stumpf (99) had already indicated that an elementary tone containing no overtones—in contrast to the tone color of the overtone spectrum—possesses a "tonal color" (*Tonfarbe*). Sine tones of low frequency have an "oo"-character, very high ones an "ee" sound and the middle range has an "ah"-character, as one can demonstrate with an electronic tone generator. Very deep tones sound "black," very high tones "white."

One can further imagine that in the physiological hearing processes described here the psyche is involved, now guiding one's attention more to pitch and now more to tone color. There is a fluid transition between the two phenomena.

In any case, the grasping of color appears primary—in physiological as well as in psychological terms; perception is formed first through a psychic process of fusion of spectral components (cf. p. 112). A prerequisite of such fusion is a certain balanced and integrated distribution of energy within the spectrum, in which, of course, concentrations in the sense of formants are possible; the spectrum must not be too strongly segmented or interrupted, since then single components of the spectrum will be isolated and will be perceived as individual tones (98). Through practice the psychic ability to concentrate can be brought to the point of recognizing individual sound components out of a rounded complex of sounds, a technique which C. Stumpf in particular had mastered.

10. The Smallest Perceivable Dynamic Fluctuations

For musical hearing it is also interesting to know how sensitive the ear is to slight variations in dynamics. In the course of the above observations it has become clear that permanent tones with constant sound pressure do not occur in speech or in music.*

When we have a value for the differential threshold of loudness fluctuations, we obtain indicators which show how inexact sound systems such as musical instruments or electroacoustic transcription systems (television, radio, recordings, etc.) may be with regard to intensity fluctuations. But one must consider how fast an amplitude change occurs, for a very sudden jump in amplitude, e.g., the switching on of an acoustical event, is heard as click noise.

The measurements are made in the following manner: A constant sine tone undergoes an amplitude modulation (Fig. 27), which is the same as a loudness fluctuation (amplitude vibrato). One finds that the ear is especially sensitive to amplitude fluctuations at a fluctuation frequency of 4 cps. With sufficient volume (80 phons) fluctuations of 1.5 per cent are noticeable. This is valid in the area of from 60 to 500 cps; above this the differentiation ability is lower. With increasing modulation frequency, the amplitude modulation is heard as roughness (e.g., 30 cps with a fundamental of 250 cps); here the ear is much less sensitive to the fluctuations.

Recalling the limit of the absolute sound pressure fluctuation which is audible, we find at a sound pressure level of 80 db,

* The organ point has only a supporting function.

corresponding to sound pressure of 2 μb, a fluctuation of ±0.04 μb, or $\frac{1}{50}$, which is just noticeable.

As measurements with sine tones have only theoretical interest, they were repeated with noise in the hope that one would discover something about the character of speech and music as a series of preponderant transients. Segments of white noise were employed. Surprisingly, the barely audible frequency shift is independent of band width and, for the most part, of sound pressure level (19). The sensitivity to fluctuation frequency is about the same as for sine tones. Thus one masters the acoustical model of speech, since vowel changes occur on the average of 4/sec intervals. The result is that a modulation degree of 6 per cent will just be perceived—corresponding to fluctuations of ±6 per cent, or the barely audible increase in intensity of 1 db. In the area of 10 to 130 db, 120 loudness levels of noise can be distinguished. Approximately this number is employed in symphonic music, whereas for commercial speech 30 levels suffice.

11. The Smallest Perceivable Pitch Differences

For musical perception the ability to differentiate very small pitch differences is quite significant, a point to which we will return in the last section of this chapter.

According to recent measurements (19), pitch fluctuations of 3 cps in the area up to c^2 are just audible; that is, about a quarter tone at *G*, $\frac{1}{20}$ of a tone at c^2. Above c^2 the pitch differentiation threshold is constant at $\frac{1}{20}$ = 10 cents, while fluctuations with a frequency of 0.6 per cent are audible. A quarter-tone piano, which for centuries has repeatedly been suggested, would accordingly only be usable from *G* up. The unclarity of the lower tones would make it unsuitable for the formation of melody. The ear is nearly twice as sensitive to pitch fluctuation at high volume levels.

According to these observations, approximately 850 different pitches can be distinguished by our ears. The ear is especially sensitive to periodic frequency fluctuations, e.g., the wow in recordings resulting from irregular revolving of the turntable. Frequency fluctuations from 0.2 to 0.3 per cent at 78 rpm = 1.3 rps are clearly heard from out of the original sound structure. The highest sensitivity occurs with the periodic change of 4/sec, as mentioned above.

The onset behavior of the hearing mechanism can be determined from its frequency resolving power according to the relations existing between the frequency band width Δf and the transient time Δt of a filter:

$$\Delta t \cdot \Delta f = 1$$

If one accepts the resolution power of the ear as 0.003, one arrives at the result (76) that the ear system requires a time of 165 periods of the incoming vibration before a stable condition is reached.* Only then will a pitch of a stationary vibration with a maximum selectivity be perceived. This theoretical value of 165 periods is not dependent upon the frequency, that is, the onset time is shorter with higher pitch. The onset behavior of the ear does not prevent tone character and the pitch from being recognized earlier, as we shall see below.

In order that unequivocal pitch perception occur, a tone in the favorable area of 1000 to 2000 cps, according to Békésy, need last only approximately 12 ms (A). It takes about four times as long, approximately 50 ms, for the significant tone color to be formed (cf. p. 112 and 118). Successive sound events occurring more quickly than this can no longer be perceived separately, for 20 impulses/sec are already perceived as a unified single tone.

After becoming familiar with the differential threshold of the ear with respect to loudness, pitch and time, we can divide the audible field into the smallest differentiable elements, taking cognizance of the differences between the application of this to sine and to noise tones (Figs. 76 and 77) (117). For sine tone application there are 720 tones with gradations of loudness from 1 db at 30 db to 0.2 db at 100 db. The greatest concentration of the squares, the maximal information content, is found between 700 and 5000 cps, especially at great loudness. The raster for noise application is less fine but more even. The concentration is greater relatively between 700–5000 cps, but does not increase with great loudness. The informational use of the speech–music area is better explained by the noise application than by that of sine tones.

This presentation is the basis for the computation of the information capacity of the ear, used in daily life.

The representation in Figure 78 forms the basis of sound analysis. If the frequency resolution, for instance, is made finer than 50 ms,

* For $\Delta f/2f = 3 \cdot 10^{-3}$, Δt becomes equal to $10^3/6f = 165\ T$, where T is the period.

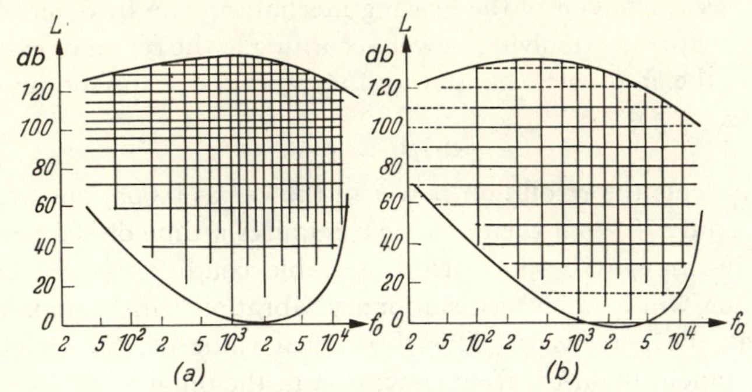

FIG. 76. Hearing area divided into zones of differentiation.

(*a*) Sine tones. One square corresponds to 1000 differentiable sine tones. (*b*) Frequency bands of colored noise extracted from white noise. One square corresponds to 100 differentiable noise characters (according to E. Zwicker).

the time resolution must be accordingly rougher, so that the resultant rectangle and the square of uncertainty will have equal area.

In reality, the resolution raster is not nearly so fine since the differentiation thresholds for all parameters are larger as far as musical processes are concerned. That is valid also for the perception of time units, since in a tone series with a velocity of 20/sec, metathesis would set in.

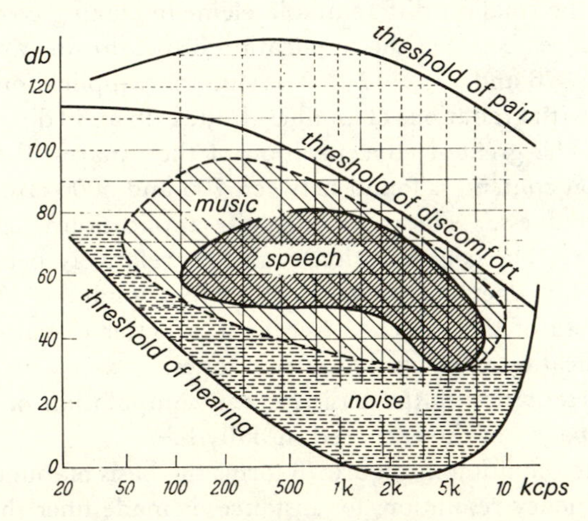

FIG. 77. Information density of the hearing area for speech and music, showing the thresholds of discomfort and pain.

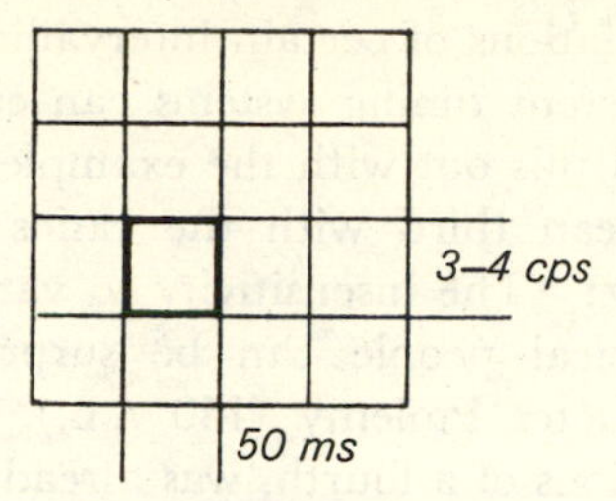

FIG. 78. Resolving power of the ear. Mean resolution 0.2 db, 3–4 cps and 50 ms.

According to a hypothesis of Potter (78), it is assumed that the image of an acoustical event in the brain occurs psychophysically in a manner analogous to the Visible Speech representation (79). It is then scanned for psychological evaluation in higher centers of perception.

12. The Uncertainty of Interval Relationships

The raster concept, or the concept of raster uncertainty, as employed in Figure 78, has rather extensive significance in science. A biological example will serve to make the problems more clear. In order for the surrounding world to make a conscious impression through the eye, visual sharpness or resolving power is important. Accordingly there is a place raster, differing in fineness according to the various types of organisms.

In contrast to place raster there is the time raster, which is expressed not only in the spontaneous movements of a living being, but also in the speed with which the eye is able to scan the field of vision (13). Here again we find a reciprocal relationship existing between the place and time determination in the fineness of the raster resolution, which in practical television, for example, derives from the above case of the frequency–time raster. Thus the expression of the rhythmic accomplishments of the protoplasm of a subject is sensibly coordinated with the relationship of place and moment (see p. 50).

The uncertainty of the analyzing capability of our sensory organs indicates that it is not too important to hold with excessive exactness to the mathematically given intervallic relationships in music, and that the slight deviations of certain intervallic relationships in the comparison of different tuning systems can easily be overstressed. Handschin pointed this out with the example of the natural third and the Pythagorean third with the ratios of 4:5 and 64:81, respectively, saying: "The insensitivity to variations in intonation even of very musical people can be surprisingly great" (32). Handschin also quotes Ptolemy (150 A.D.), who, in his investigations of his divisions of a fourth, was already of the opinion that divisions which differ by tone deviations of only $\frac{1}{24}$ of a whole tone (8.3 cents) are superfluous as individual genera, as is also the distinction between the two types of whole tones 9/8 and 10/9 with 22 cents.

Further, Hindemith (34) gives the example of the five different types of thirds: the major 4:5, the minor 5:6, the "too small" 6:7, the "too large" 7:9 and one lying between the major and the minor 9:11, finally also the Pythagorean third, which lies between 4:5 and 7:9. All these intervals are recognized by the ear simply as a third (see Table II). In connection with this there is the following experiment: using, for example, a violin, produce the smallest interval greater than a second which is still perceived as a third, and move the upper tone up towards the fourth, to that limit

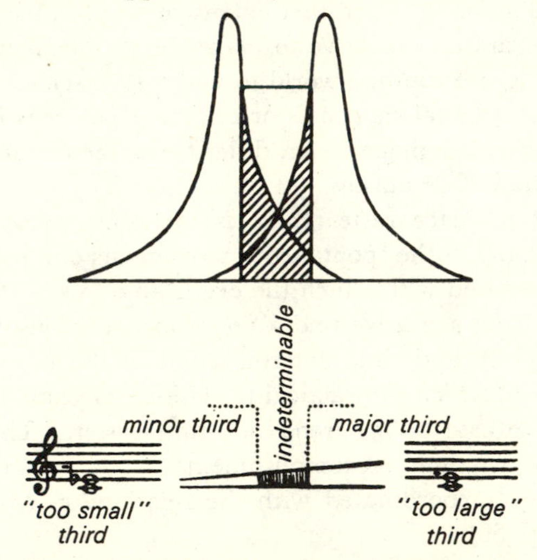

FIG. 79. The uncertainty of intervallic perception.
Above: response width of the sensory pick-up system of the basilar membrane.

TABLE II

Third Relationships, calculated in accordance with $c^2 = 500$ cps

"Too small" third	7:6	585
Minor third	6:5	600
Third between major and minor	11:9	611
Major third	5:4	625
Pythagorean third	81:64	633
"Too large" third	9:7	645

where one still has the impression of a third; during this gliding of the interval one cannot establish precisely where the separation exists between major and minor third. The sensitivity curves dovetail. In the middle of the third there is an area which can belong to both types of third. The ear under certain circumstances can recognize the at first uncertain interval as a major or minor third from the context of the melody or harmony (Fig. 79).

Chapter VII

UNCLARITY IN MUSICAL STRUCTURES

Perfectly correct music cannot even be imagined, let alone executed.—Schopenhauer.

1. Fundamentals

In the following the attempt will be made to determine to what extent the practical performance of a piece of music corresponds to the prescribed note values. Such an investigation is important, because even through relatively small deviations in intonation the performance no longer maintains the strict intervallic relationships in the sense of harmonic theory, as the composer imagined while writing down the note values. In this we understand by deviations in intonation not only those caused by a performer's accidental inaccuracies in pitch, but also the conscious as well as unconscious deviations from the "target" tone and ornamental embroideries of it. Finally, those foreign tones will be investigated which occur when several tones are sounded together, to see whether they correspond to the strict principles of intervallic theory.

The observations in this chapter are limited to stable or quasi-stable conditions of sound, because the onset transients—as discussed in the first part of the book—proceed from an as yet quite undefined formation in tone perception, out of which unambiguous tone perception is crystallized only afterwards.

Our task is facilitated by the existence of various recording devices that make it possible to read or to compute the deviations from the ideal values.

The problem arises in the following situation: In the quest for greater accuracy in intonation on the part of the performing musicians, and greater purity in sound and constancy on the part of the builders of musical instruments, the goal has been attained in the field of electroacoustical instruments, which possess nearly perfect accuracy in the presentation of the desired note values. The surprising result is that such idealized tones, in spite of all enrichments in color, are dull and unexciting. Is a certain inaccuracy in intonation actually necessary for a satisfying auditory impression, and how great may this inaccuracy be?

2. Intonation Inaccuracies in Practice

Various investigators have made measurements of the accuracy with which written note values are performed by instrumentalists and by singers.* Of course only first-class artists were employed.

As far as singing is concerned, C. E. Seashore (89) has communicated that the professional singer never reaches the desired pitch in 25 per cent of the tones, whereas in the other 75 per cent the exact pitch is only momentarily reached during the time span of the given note value.

In the case of a tone sung after an organ tone Sokolowsky observed, in statistical experiments, a discrepancy of ± 0.8 per cent, while in the free singing of intervals, there were considerably larger discrepancies, e.g., in the case of a fifth, 3.3 per cent. The deviations increase with fluctuating dynamics. The constant fluctuation in dynamics is greater in the human voice than in instruments. It is interesting also to observe that the pitch is more accurate in singing *legato* than in singing *staccato*, where deviations as large as a half and even a whole tone occur without being particularly noticeable to the ear. One can see from Figure 80 that *staccato* attacks last on the average 0.1 sec, and even in this short time very considerable frequency fluctuations occur (27). Because of onset and other transients, which, according to our physical explanations, lead to a noisy smear, the distuning remains essentially hidden from the ear.

Helmholtz had already investigated in detail impurity of singing. Also E. M. von Hornbostel established that Europeans sing "terribly inaccurately." O. Abraham drew attention to the fact that intonation in unaccompanied purely melodic song is certainly not without its rules. Our interval characters are established only according to magnitude, while the actual step sizes are determined

* Cf. also the fluctuations which arise from the onset transients (p. 31f.).

in each case by the melodic form: the leading-tone step is generally made smaller, large leaps—especially ascending—frequently exceed the target tone, a series of steps in the same direction tends to be spaced equidistantly. Abraham, too, speaks of the "enormous deviation in intonation" even with very musical singers, which are most noticeable in dissonant intervals, especially the half step. It is significant that it is not the isolated intervallic step as an absolute magnitude which is important here. The deviations are generally so great that they generally extend beyond the difference between absolute and tempered tuning, e.g., in the case of the third.

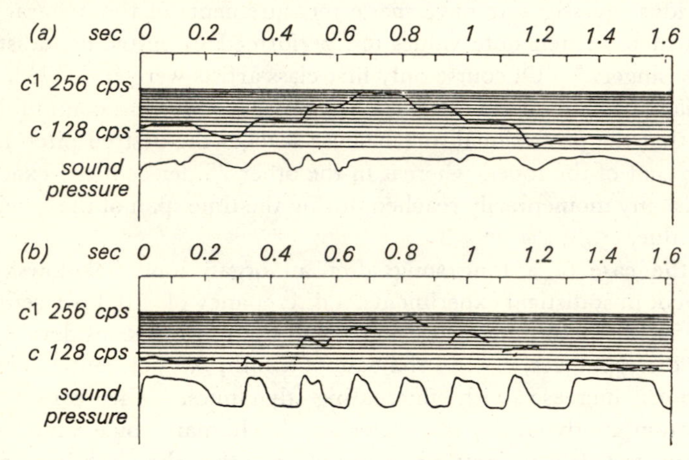

FIG. 80. Intonation of the tone series *b*, *d*♯, *f*♯, *b*.
(*a*) *Legato;* (*b*) *staccato*. Upper curves in *a* and *b* show pitch fluctuation; lower curves show dynamic flow.

L. Euler made the notable observation (1764) that the ear usually hears what it wants to hear, even if that does not correspond to the acoustically given interval—which is indeed to great purpose, since with absolutely pure intonation, which never occurs in practice, the enjoyment of music would be impossible. His statement that the acoustically exact realization of the written music would increase the listener's pleasure seems questionable. This is precisely the statement which this book attempts to disprove. Let us now look at intonation in instrumental music.

J. F. Nickerson (71) has reported on an informative experiment in which Haydn's "Emperor" Quartet was selected for comparison of the intonation accuracy of six different string quartets. The music was recorded on 16-mm film and processed stroboscopically. The determination of the deviations as opposed to the different tuning systems yielded a minimum with reference to the Pythagorean tuning and a maximum with reference to pure intonation. The deviations from equal temperament, especially in the case of thirds and sixths, showed that the effect of the omnipresent orientation to the tempered tuning of the modern piano on a performance without piano is negligible. Accordingly, Nickerson ascribes little significance to pure tuning in the framework of the accuracy of intonation in the performance of music.

Let us remember finally that even in the case of the tuning of instruments the absolute value is seldom achieved. It has been observed that the tuning of a grand piano by a good piano tuner normally reflects deviation up to 10 cents from the mathematically correct equal temperament. In addition to this, a short time after the tuning there are the seasonal fluctuations of temperature, dampness and so forth.

In examining more closely the construction of keyboard instruments, we find that in the simultaneous attack of three strings in a unison the chances of purity of intonation have already been reduced. On the average, of course, the intonation can be held, but because of the small intonation deviations among strings in a unison vibrations are aroused which will be discussed in more detail in Chapter VIII. It also occurs that the piano tuner intentionally permits a minimum fluctuation within each unison of strings. The result is certainly an "indistinctness" of the sound, but at the same time a greater stability in the physiological sense of hearing (cf. p. 161).

How impurely an orchestra plays together, or must play together, can be seen in the investigations of Lottermoser and Braunmühl (55). With an apparatus for registering pitch the consciously and carefully prepared intonation of an internationally prominent orchestra was recorded at the exact tuning tone of a tone generator. The deviations from $a^1 = 440$ cps can be seen in Figure 81. The second violin and viola have tuned to 442, the oboe only to 438, the trumpet is as much as 4 cps ($\frac{1}{6}$ of a half tone) too high. Afterwards, they played the *Tragic Overture* by Brahms, in which the tone a^1 often occurs. We can see in Figure 82 what large frequency fluctuations in the sound of the orchestra occur (up to ± 5 cps). After this piece, tests were made to see the extent to which the tuning

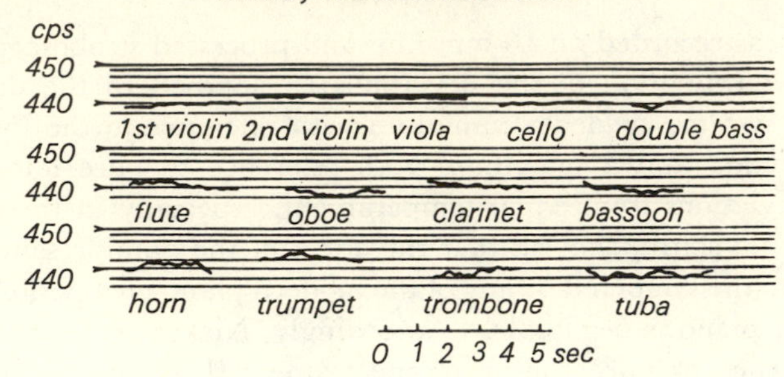

FIG. 81. Measured tuning of orchestral instruments shortly after tuning to $a^1 = 440$ cps.

of the instruments had changed during playing (Fig. 83). The strings had approximately maintained their pitch level, the winds fluctuated; the clarinet was 4 cps low, the tuba 4 cps high. The raising of the pitch is due to the warming of the instruments during playing.

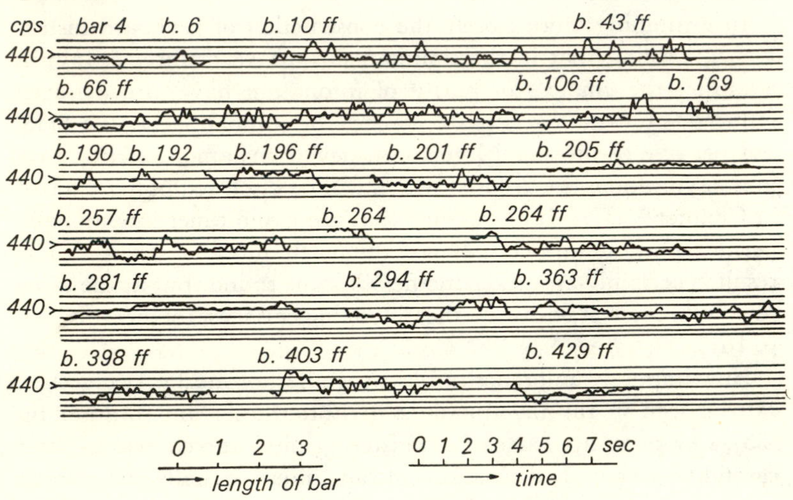

FIG. 82. Recording of the tone a^1, selected from Brahms' *Tragic Overture*.

Distuning, however, can occur in quite another way. Acoustical deficiencies—e.g., a decrease in volume caused by resonance peaks occurring in an instrument at a particular point in the tone scale—will be more or less consciously equalized through distuning, or

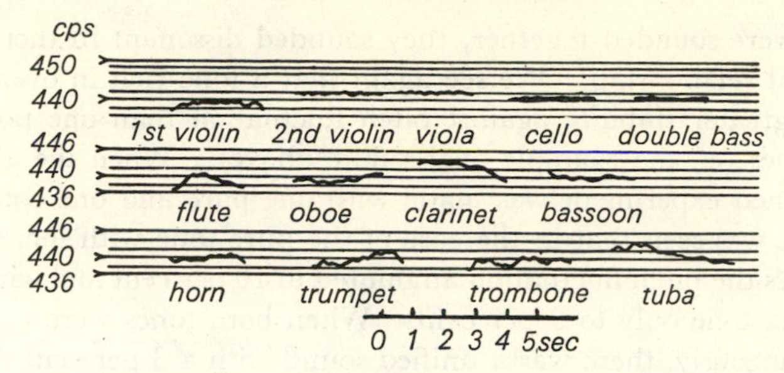

FIG. 83. Recording of the tone a^1 at the end of the concert.

even through an increase in vibrato. A singer who cannot make himself heard over the orchestra in a unison passage will make use of a slightly higher tuning, more or less consciously.

3. Intonation Deviation as a Function of Loudness

The exactness of intonation of the quasi-stationary sound of string instruments depends very much on bow pressure (cf. also p. 34). If, for instance, the open *a*-string is bowed with medium pressure and the exact pitch results, a frequency fluctuation of 1 to 1.5 per cent from this value will result with increasing bow pressure (29). Similarly, an intonation fluctuation can be observed in wind instruments with increasing air pressure. In the case of the recorder 2.5 per cent fluctuations have been registered and 1 per cent for the clarinet. Grützmacher and Lottermoser thus warn against employing reed instruments, such as the clarinet or oboe, as tuning norms (29). In the cross-blown flute frequency fluctuations up to 4 per cent have been observed. Although completely pure intonation is most difficult on the flute, it is also of least importance, for the tone is poor in overtone content, and a slight distuning can have a positive equalizing effect. One can assume a proportionality between attack strength and frequency fluctuation for all instruments.

A psychophysical discussion of such behavior has already been given (p. 93). H. Fletcher (22) has shown experimentally that a sound of certain loudness *appears* deeper if the loudness is considerably increased. The deviations can extend to a whole tone. In one experiment he intoned alternately two overtone-free tones of 168 and 318 cps with such great loudness that they sounded at the interval of an octave like 150 and 300 cps. However, when these tones were sounded together, they sounded dissonant in their real physical relationship. We see again that a tone rich in overtones has a greater stability against pitch fluctuation than one poor in overtones (cf. the example of the flute above). When the above-mentioned experiment was made with one pure and one complex tone, it was seen that in the case of the pure tone with increasing loudness the pitch fluctuation amounted to 10 per cent and with the complex tone only to 3 per cent. When both tones were intoned simultaneously, there was a unified sound with a 3 per cent deviation. Composers and conductors must therefore be aware of the fact that the effect of a distuning can occur under certain circumstances with a *crescendo* to a *fortissimo* in solo voices without the players' being at fault. Frequently there are arguments between singing teacher and pupil in this respect.

4. The Modulation Effect for Sound Expression

If we occupy ourselves further with the quasi-stationary segment of an intoned instrumental sound or vocal sound—expressed by a single note—we will see in the sound notation, e.g., in a Visible Speech picture, that the sound is at work and constantly changes (Fig. 38). These changes are partly accounted for by the modulatory characteristics of the instrument and the abilities of the player.

If we examine the intoned sound of unchanged pitch and loudness more carefully, we see an inner modulation even without the addition of a color change or vibrato. If, instead of the Visible-Speech presentation (Fig. 38), we take the vibration curves of the individual overtones of a sound (oscillogram) and arrange them one above the other, as we have done for the organ pipe (Fig. 39), we will see that the individual overtone groups which occur in octaves throughout the frequency area do not have the even character of sine waves (Fig. 84). The two lower curves are periodically damped vibrations, in each case beginning abruptly and continuously decreasing. The upper curve of the whole sound represents a superposition of all partials present, even those that are not depicted separately.

According to the information thus far, we regard every damped train of waves as a transient process, which signifies a widening of the frequency band (Fig. 19). The oscillogram shown accordingly has more color noise than one with pure sine waves.

Physiologically, we see in the representation of a singing voice the way in which the periodic push from the glottis in the larynx brings about damped vibrations in the damping resonance cavities of the vocal tract. For the nature of such vibrations F. Trautwein developed a push-formant theory, which he has demonstrated by means of an electric circuit (101).

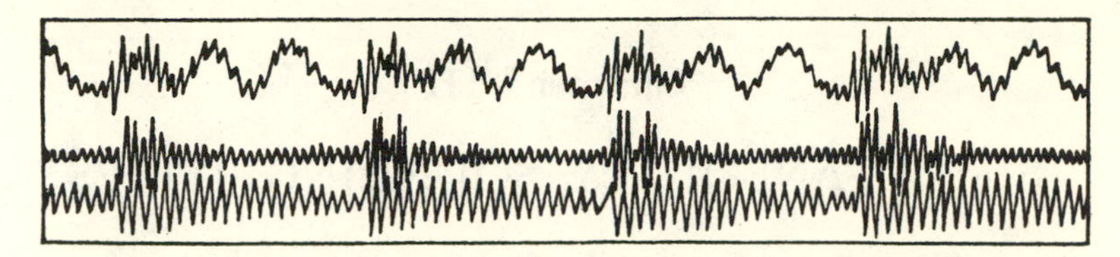

FIG. 84. Oscillogram of the vowel "e," (above).
Fundamental tone 100 cps; 1st and second overtones in the filter bands 1200–2400 cps and 3200–6400 cps, respectively.

In addition to the color variations of an "inner modulation," we also have such frequency modulation as vibrato, tremolo, and trill, etc. Throughout the history of music the techniques of composition, improvisation and performance practice have demonstrated the importance of a modulatory influence on sound in order to achieve a musical value, and recently this has been recognized in experiments with electronic music. "Hairbreadth" accuracy without any fluctuation and without vibrato is perceived as dull or mechanical.

The inner modulation of sound can be particularly variable in the human voice. It is used in long-held tones (*sostenuto*) when, for example, an "ah" is changed phonetically into "ao" or the reverse, or a vowel is gradually brightened through an articulatory "switching-in" of the "ee"-formant or darkened through the addition of the "u"-formant. This inner modulation, which is a spectral modulation—change of the overtone spectrum through transients—and is guided by the inner supporting functions of the human body and a subtle articulatory motor mechanism, creates a dynamic of complex tone colors in the place of that other means of expression, the ornamentation technique of Baroque instrumental and vocal music. This technique apparently was used because of lack of inner control in influencing the sound (cf., for example, the significance of the *Pralltriller* for long-held tones when they are intoned by the human voice or on the piano).

Chapter VIII

SIMULTANEOUSLY SOUNDING TONES

Up to this point we have observed the vibration behavior of a single tone and then also that of an individual sound, and we have attempted to describe the psychophysical phenomena which are the basis of the acoustical stimulation effect on the inner ear.

What phenomena are to be observed further when any two tones or sounds strike the ear simultaneously? First of all we must establish that in music seldom, if ever, does a single tone occur by itself, since as a result of the decay of the instrumental sound and the resonance of the room, every tone is accompanied by the tone preceding it (pedal effect).

1. Beats

According to the law of superposition, two tones sounding simultaneously are superimposed in a very simple way: one adds their amplitudes (Fig. 27). If the two tones are of nearly equal frequency—e.g., two members of a unison of piano strings—they blend into a single tone and one hears beats, that is, loudness fluctuations with a periodicity which is in the frequency of the difference between the two original frequencies. A distuning of 3 cps, for example, is clearly recognizable as a fluctuation of the resulting tone with a period of three times per second. Faster beats, e.g., 30 cps, appear to the ear accompanied by a roughness in tone. It was possible to find the beat flow as a synchronous impulse pattern in the nervous system, but we cannot say in which part of the nervous system the formation of the pattern occurs (83).

Still, slow beats do not play a significant rôle in Occidental music, since tone changes generally occur much too quickly and transients and fluctuations of other sorts do not permit the stationary vibration of two tones. If the audible impression of a periodic amplitude modulation occurs because of beats, it is not unpleasant for the ear, as is shown, for example, by the experiments with vibrato which were produced with an approximation of amplitude modulation of 6–7 beats per second. These amplitude changes from period to period are favorable. From the standpoint of perception of musical tones, the beat and not the simple sine tone can be considered as the simplest building block of musical texture, as we have mentioned above.

The audible impression of beats is dependent upon pitch. While in the contra-octave (32–64 cps) the intervallic distance of a major third (4:5) is still perceived as a beat, this effect in the octave above 2000 cps is not even possible for the minor second (17:18).

Ranke, Husson and others have drawn attention to the fact that beats do not only consist in a loudness fluctuation but also in a fluctuation of the resulting tone between the pitches of the two initial tones as long as they have equal amplitudes (36, 81). We can see in Figure 53 that with great sound intensity several nerve fibres of the hearing nerves are excited, and with less intensity only a few. The tone that sounds first (in this case, e.g., 500 cps) initiates the stimulation, whereby the corresponding fibre is no longer excitable for the duration of the refraction time ($\frac{1}{1000}$ sec), so that the second tone, which occurs within the time span of $\frac{1}{500} \times \frac{1}{20}$ sec, is not able to yield a primary stimulation. Here it is assumed that both neighboring tones are in the influence area of the same frequency group of hair cells in the inner ear. With the next swelling of the beat tone it is the deeper tone (480 cps) that occurs first, as a result of the inversion of phase in the beat minimum. This fluctuation results in a lowering of the pitch sensation, which is considered of positive value as a fluctuation phenomenon in music.

2. Combination tones

Difference tones are not to be confused with beats. Difference tones result from a distortion effect of the reproduction system.

This can occur, for instance, in the acoustical generators of instruments (membrane, soundboard, etc.) and in amplifiers and loudspeakers in the case of electronic reproduction systems. To characterize these systems, one speaks of "non-linear distortion." Sum and difference tones result, which together are referred to as "combination tones." These were discovered as early as 1740 by the organist W. A. Sorge, who, on playing two organ sounds together, heard a third, deeper tone. Sum tones often lie so high that they are scarcely heard. More important, however, are the difference tones, which contribute significantly to the construction of a sound, supplementing or reinforcing the foundation of the deeper tones, which is desirable in the case of insufficient scanning of the lows of instruments and loudspeakers. As we mentioned above (p. 56), the low frequency onset transients give a sound more form and depth and the effect of being nearer to the listener. If we generate a constant tone in the tone generator of a loudspeaker, and sing along, first in unison, with the same tone, and then at different intervals, we will hear quite clearly additional deep tones. From this we can conclude that the additional tones have originated in the ear itself.

From these observations we can draw up the law for the formation of the combination tones f_k for two original tones with the frequencies f_1 and f_2:

$$f_k = m \cdot f_1 \pm n \cdot f_2$$

in which m and n represent the series of all whole numbers. Through experimentation it has been discovered that as many as 76 combination tones of a measurable intensity can be perceived by a cat.

The tuning of the violin in fifths is therefore particularly simple, since with this interval relationship, e.g., 300:200 cps, the suboctave with 100 cps results as a difference tone. Thus one tunes with the first three partials of the overtone series. The ear is particularly sensitive to distortions of the fifth.

The sounding of an individual tone of the overtone structure can also produce difference tones because its partials permit the formation of sum and difference tones. In an overtone series of 100, 200, 400, . . . cps the difference in each case between partials is 100 cps, resulting in a reinforcement of the fundamental. This is simple to prove experimentally by filtering out the fundamental; because of the formation of difference tones the fundamental is still heard. Organ builders make use of this knowledge for the formation of very low tones for which they would otherwise have to build especially large pipes; the tones are generated through difference tones produced from two smaller pipes. Oddly enough, the fundamental is also supplied by the ear, when there is compensation for the combination tones which result from non-linearity. The formation occurs by means of a number of higher harmonics, the "residuum," which absolutely must remain if the effect of the fundamental is not to disappear (86).

Further, one must note, along with O. F. Ranke, that the nonlinear characteristic of the ear may be an individual matter (80). For example, when the distortion in the middle ear of some subjects is especially strong, the combination tones will be heard louder than normal. People who so react would be the best piano tuners. The same music can be heard quite differently by different people. C. Stumpf has further observed that the difference tones are perceived quite a bit louder than they should be theoretically, according to physical computation. Here the residuum tones perhaps play a rôle. Skudrzyk explains the greater loudness as an analogue to the optical distance distortion in projection (L).

3. The Subjective Character of Intervals

In order to understand better the behavior of subjective combination tones, that is, those formed in the ear or, in other words, those that are not present objectively in the air, we will make use of the classical diagrams which Hindemith also discussed (34) (Figs. 85–87). To a fundamental c^1 (horizontal line) tones of the chromatic scale are sounded in ascending order. In the case of c^1 with c^1 (unison) the difference is, of course, zero; however, the octave c^1 with c^2 yields c^1, and the fundamental will thus be reinforced. The complete scale for two tones between c^1 and c^2 yields combination tones which behave according to the drawn curve. As can be seen in the case of the major third and also the fifth, the lower octaves of the fundamentals in both these cases are supplied. In this way the lowest tones of many instruments which can no longer be projected by the resonators, are automatically supplied (e.g. violin, grand piano, organ, human voice, etc.). The musical notations accompanying the diagrams show further that the other scale steps are also harmonically supplied. The combination tone is really problematic only when intonation is not pure, for then it deviates from the simple integral intervallic relationship and brings into prominence the psychic invitation of out-of-tuneness. The curve is not closed at the bottom, because below the first interval the difference tone loses its effect as a coherent tone.

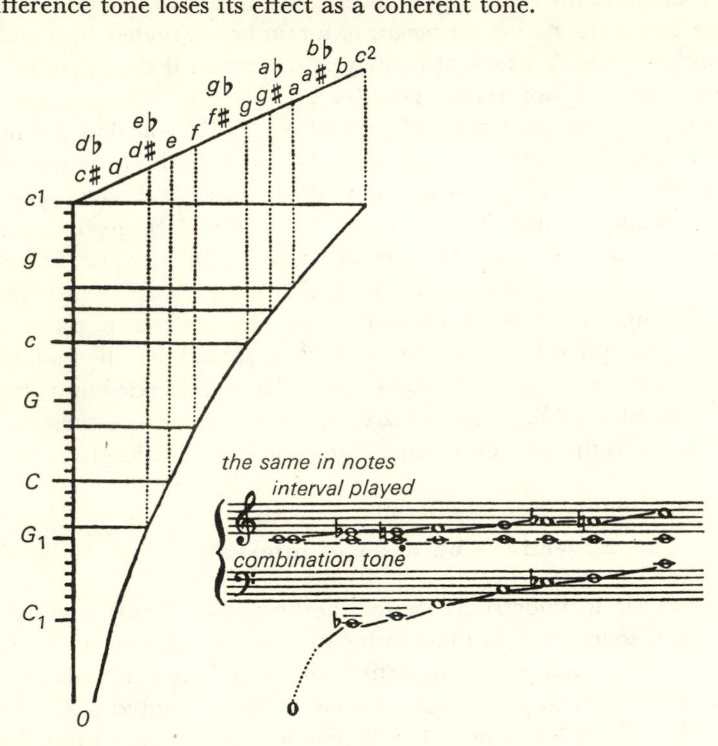

Fig. 85. Representation of combination tones of the first order (according to F. Krueger, 1903).

Since combination tones appear as independent new tones, they themselves form new combination tones with the original tones; these new combination tones are called combination tones of the second order, because they are not as loud as combination tones of the first order (Fig. 86). The computation of the curve is a result of the formation of difference tones in each case between one original tone of the scale and one combination tone of the first order. One could theoretically form new combination tones of the third order, and even higher, with these new tones, but these are scarcely audible (Fig. 87).

As is shown in Figure 87, unison and octave sound without any addition; in the fifth, combination tones of the first and second order coincide, whereas all the other intervals have a double burden of varying weight. In these cases the combination tones give the intervals a "personality." "An interval without combination tones

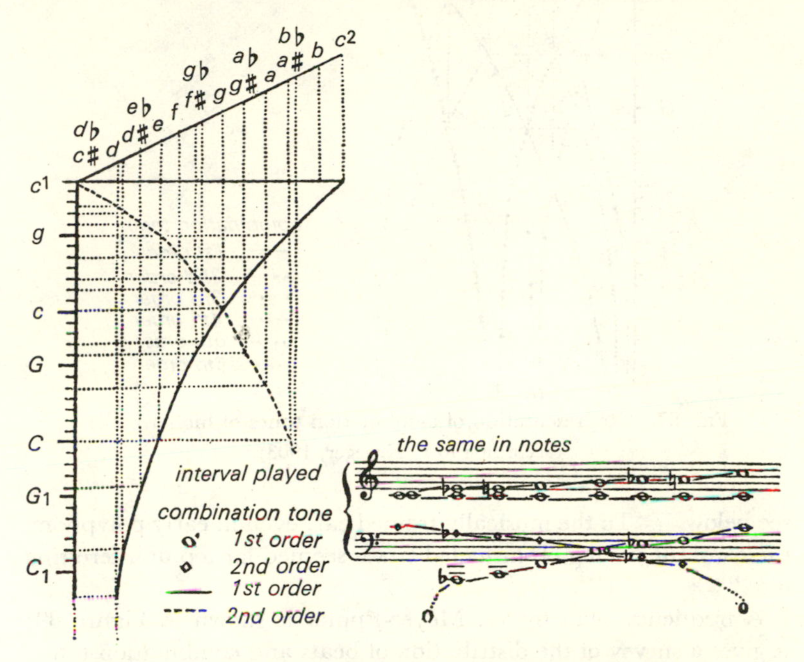

Fig. 86. **Representation of combination tones of the second order (according to F. Krueger, 1903).**

would be an abstract concept without being" (Hindemith). Here it can be remarkably clearly observed that the harmonic components in acoustical spectral lines are of little significance and that it is precisely the "indistinctness" that gives the sound its form and content, making it fit for use in musical construction and development. With this viewpoint Hindemith regards the minor triad as an "indistinct" derived form of the major triad (compare this with the combination tone pictures of major and minor chords, Fig. 88). Both chords are not clearly separated from one another, as we will

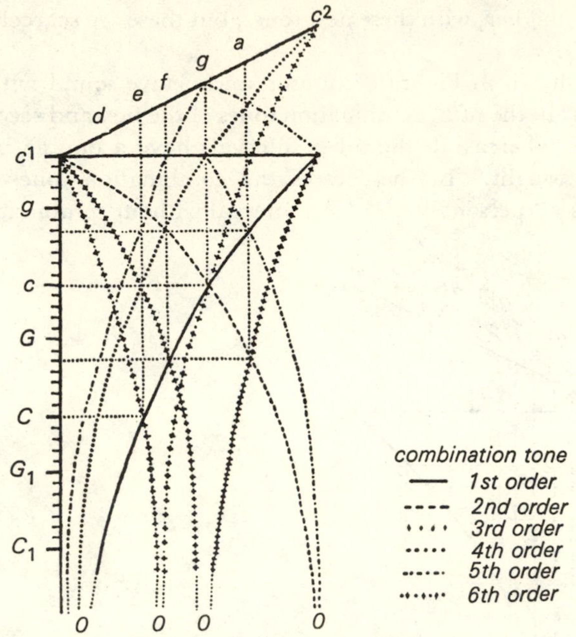

FIG. 87. Representation of combination tones of higher order (according to F. Krueger, 1903).

see below. "To the musically trained ear, even in early polyphony the constant joining of unshaded triads seemed far too uninteresting fare."*

A frequency plan by W. Meyer-Eppler is shown in Figure 89; it gives a survey of the distribution of beats and combination tones, employing a fixed tone of 200 cps against a second tone varying

FIG. 88. Behavior of the combination tones in a major chord (LEFT) and in a minor chord (RIGHT).

* "Dem musikverständigen Ohr sind schon in der frühen Mehrstimmigkeit die unentwegten Verbindungen von ungetrübten Dreiklängen als gar zu reizlose Kost erschienen."

from zero to 400 cps (65). The heaviness of the lines represents the relative strength of the combination tones.

According to Helmholtz, we can derive the sensation of consonance and dissonance from this behavior, speaking of dissonance

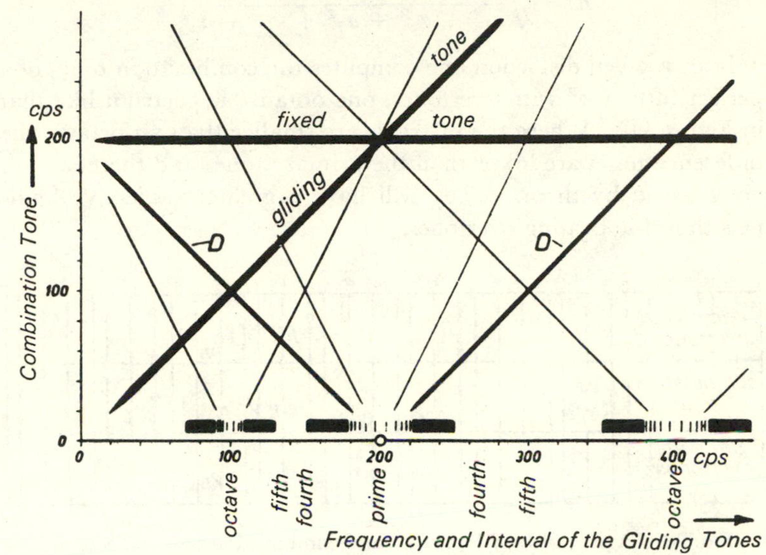

FIG. 89. Frequency plan of the combination tones resulting from simultaneous intonation of two tones of different pitch.

D = primary difference tone; ||||| = beats; ▬ = acoustical roughness.

when there is an interval between at least two tones in which roughness is audible. C. Stumpf, however, has said that an interval becomes increasingly consonant as more fusion of two tones is perceived. This is doubtless especially true in the case of the octave.

4. Measurement of the Audibility of Distortion

W. Weitbrecht (107) has conducted extended experiments with combination tones of the first order, in order to measure quantitatively the much disputed audibility of these additional tones. The experiments were conducted on an electroacoustical basis, the tones being distorted by rectifier tubes and the resultant so-called non-linear distortion being measured electrically as the factor K

of non-linear distortion (*Klirrfaktor*). This is the sum of the overtones caused by the distorting system a_2, a_3, etc. in comparison with the fundamental a_1 plus these overtones:

$$K = \sqrt{\frac{a_2^2 + a_3^2 + \cdots + a_n^2}{a_1^2 + a_2^2 + a_3^2 + \cdots + a_n^2}}$$

If for a given distortion one computes the combination tones of a perfect fifth c^2–g^2 with overtones, one obtains a spectrum like that in Figure 90. Where the intervals are smaller than an octave, the difference tones are lower than the primary tones and therefore are not masked by them. They will have a greater subjective loudness than the beating overtones.

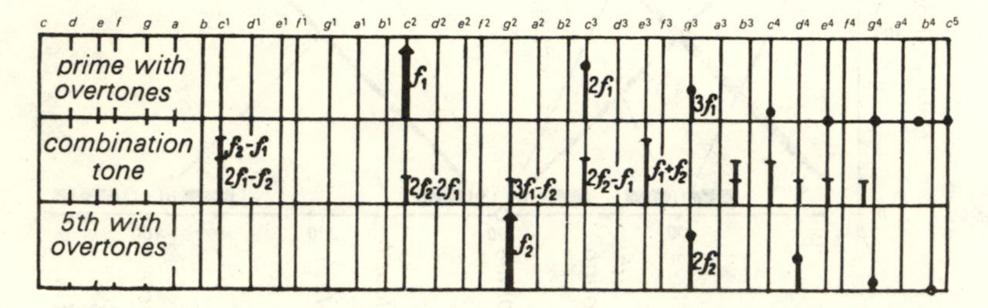

FIG. 90. A pure fifth with partials and combination tones (spectrum).

It is of further significance that the beats of greater amplitude—e.g., in a distuned octave—cause a change in the sound character also. At the beat maximum the sound is very "fundamental," and at the beat minimum one hears a bright sound rich in higher overtones. The explanation for this lies in the fact that at the moment of maximum "fundamental beat" there is a lower percentage of harmonics than at the moment of the minimum.

Here we see quantitatively how large the undertone c^1 is that is formed from f_2–f_1 and $2f_1$–f_2. The simple computation of combination tones solely from fundamentals is therefore not sufficient, as this example with the first overtone $2f_1$ shows. One can also see that even in the case of fifths difference tones are formed from the overtones which deviate significantly from the simple harmonic number relationship: b^3, c^4, d^4, e^4, g^4 impure.

If the upper tone of an interval is slightly distuned, a number of difference and partial tones occur, which lie beneath the beating overtone pair but fluctuate in the same rhythm.

We can see from this that the type or clarity of the beating is important in order that the distuning in pitch be noticed. The results of experiments by Weitbrecht are valid for an observation time of 1 sec for the sounding of the interval, whereas in practical music note values are usually of shorter duration. In order to become acquainted with the basic behavior of distuning as a natural phenomenon, we will present results derived from electroacoustical reproduction systems. At least we can see here that distuning of less than 1 cps can disturb the sound without the counting out of beat periods. The experiments were so conducted that for the purpose of comparison the listener heard distorted and undistorted tones of differing loudnesses $L_1 = 54$, $L_2 = 63$, $L_3 = 71$ phons. The listener then had to decide whether this or that sound timbre was distorted or undistorted.

Figures 91 and 92 present curves which plot the limit of distortion,

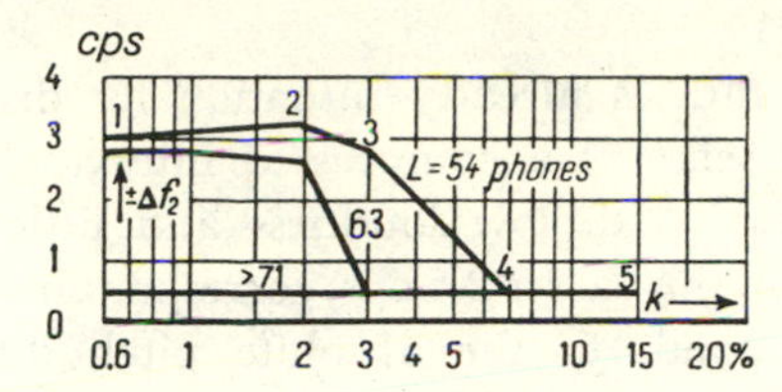

FIG. 91. Curves of the just audible distortion of a pure fifth.

that is, that distortion which is barely recognized as such according to its quality of sound. It is interesting to compare the pure and the tempered fifth. Using a medium basic loudness of $L = 54$ phons (upper line) and the least distortion, which, for example, is caused by characteristics of the ear (factor of non-linear distortion = 0.6 per cent), the distuning of the upper tone of the pure fifth from 3 cps on is recognized as a change in sound quality (point 1). If the distortion is increased—in the experiment through electro-acoustical means—to 3 per cent, the sound impression with the

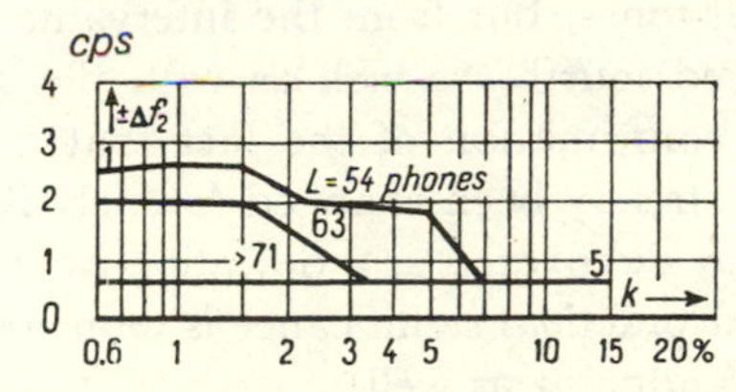

FIG. 92. Curves of the just audible distortion of a tempered fifth.

same distuning of 3 cps (the same as 6 cents) is hardly changed. Only with still stronger distortion, which does not occur in the ear but in loudspeakers, is distuning noticeable at less than 1 cps (point 4). Here we must say that up to now a factor of non-linear distortion of 3 per cent, scarcely noticeable by the ear, has been deemed adequate for electroacoustical transmission equipment. Speech remains completely understandable even with a factor of non-linear distortion up to 10 per cent. We see further in the diagram (middle line) that with greater loudness, $L = 63$ phons, the distuning is recognized slightly earlier when the distortion is small, but the limit of distuning of $\frac{1}{2}$ cps is recognized with a factor of non-linear distortion of 3 per cent. In the case of still greater loudness, $L = 71$ phons, with the very slight factor of non-linear distortion of 0.6 per cent (ear) an intonation deviation of $\frac{1}{2}$ cps is noticeable—to be sure, only in an immediate comparison with undistorted sound.

The tempered fifth is already intentionally distuned by 0.02 half tone—in the sense of a compressed interval. The beat frequency is 1.7 cps. With low loudness and little distortion the limit of distortion in this case is 2.5 cps, corresponding to 3.2 per cent. This value corresponds to the absolute pitch differentiation of individual tones (see also p. 120), so that the change of the upper tone is judged independently of the other interval components. The recognition limits of the slightest distuning are generally similar to those of the pure fifth, except that the lowest limit of recognition (line 5) is simply moved up in the case of a tempered fifth (0.8 cps) as opposed to the pure fifth with 0.5. We see from these investigations that *loudness is significant in recognizing deviation in intonation.* For musical purposes we simply have to observe point 1.

It is astounding that for the pure fourth the limit of distortion is much lower, namely 63 phons—smallest distortion 0.7—only 1.4 cps or 1.5 per cent; the change is therefore not recognized directly from the individual tones, but from the intervallic effect. In the case of the tempered fourth, as well as with the other tempered intervals, we find confirmation of the fact that a change in the interval which has already been changed (namely through tempering) is not as quickly discovered as a deviation from the pure interval. Of what great practical significance is tempered tuning when viewed from this standpoint as well!

Since beating in the tempered fourth and first occurs very slowly because of the very slight correction, it is not noticed in a fast melodic progression.

In the case of small intervals, such as the major third, it is at first disturbing that the grumbling, low-lying difference tone does not easily permit recognition of the deviation, but this effect is largely compensated by the reciprocal distuning of the various difference tones.

The same is also found in the minor third and minor sixth. Where L is greater than 98 phons a distuning of 0.3 cps is clearly perceived. The case of the tempered major third is particularly bad, since the fundamental difference tones f_2-f_1 and also $2f_2-2f_1$ are no longer consonant. (Helmholtz: "This is a hideous bass, all the more hideous because it comes rather close to the right bass.")

Much less sensitive is the minor third, in which—with low-level loudness and distortion—the distuning must be made greater than 3 cps in order to be heard. Here the grumbling of the low difference tone f_2-f_1 is again disturbing. When the primary tones are very high (soprano voice) the low difference tones can be particularly prominent. Small deviations are multiplied in the difference tone. A duet of soprano voices with tremolo may then become unbearable.

As already indicated, the recognition of the theoretically presented distortions through deviations in intonation is made difficult in practical music because of various phenomena:

1. The duration of a tone in a musical work is very seldom as long as that employed in the experiment, 1 sec.

2. The changed tone component can be masked through combination tones; here it must be remembered that the masking is greater with a low loudness level, and also that the masking is dependent upon the relative position of the masking tone to the masked tone. At 50 phons perceptibility is equal to the still barely perceptible pitch change (therefore without interval).

We can see from Figure 93 how much of the sum and difference tones caused by distortion is not masked by the original tones A and B of the interval and is thus finally audible; the masking curve is indicated with a dot-dash line. It can be noticed that only the difference tones D_2 and D_{3B} can be audible. With a decrease in consonance the beats lose significance, while at the same time the difference tones gain in significance (pure minor sixth, pure major and minor third).

The studies of Weitbrecht were continued by G. Haar (30), who investigated more thoroughly the nature of the formation of the distortion. He shows, using the example of the distuned fifth and the minor sixth, that as the frequency increases the difference tones, especially in the area of greater ear sensitivity, above 1000 cps, are perceived as more and more disturbing, therefore that the limit factor of non-linear distortion becomes continually smaller. With high loudness level and high frequencies, distortions of a few pro mille are already audible. With extreme loudness—above 100 phons—the ear creates increased distortion, which for tones around 200 cps provides a non-linear distortion factor of up to 8 per cent.

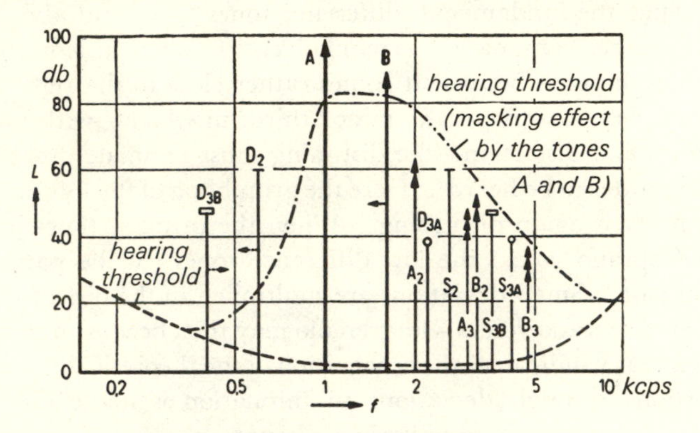

FIG. 93. Sum and difference tones resulting from the two tones A and B by a non-linearity process.

The sum tones are completely masked by the generating tones A and B, as can be seen in the dot–dash curve (according to Zwicker).

It must be noticed further that in the case of loudspeaker reproduction the line-current hum of 60 cps can form sum and difference tones in combination with a high tone, for example that of a violin; these may lie close to the violin tone and cause a roughness in it, in addition to the noise of the line-current hum. In further experiments with the violin it was established that through a vibrato, with the same limit factors of non-linear distortion, additional noises occur, with frequency ranges above the fundamental. A further response disturbance occurs because of the electro-acoustical transmission when series of rapid tones are played on the string.

This disturbance can become a factor of non-linear distortion greater than 10 per cent, but it decays rapidly. According to the formulation of G. Haar, passages on the *G*-string sound rather like "veh" and change in the direction of a nasal sawing sound as they go on the *D* and the *A* to the *E*-string. Very little remains of a musical sound substance. The origin of this distortion is explained by combination tones formed by the reverberating previous tone with the following tone. Since the original music was not affected by this disturbance, a confusion with the onset transients described on p. 34 is not possible. The same effect is at the root of the unpleasant results of an overused pedal.

It is especially important that the experiments were mostly made with overtone-free interval tones. R. Feldtkeller has pointed out that sensitivity to non-linear distortion in musical intervals becomes less when overtone-rich tones rather than sine tones are employed (20). If we assume that a loudspeaker system has a factor of non-linear distortion of 10 per cent, then, as in Figure 94, where the distortion threshold is drawn on the audible field, the so-called second harmonic distortion* k_2 of a chord with the use of sine tones and a loudness greater than 20 db would be noticed; on the

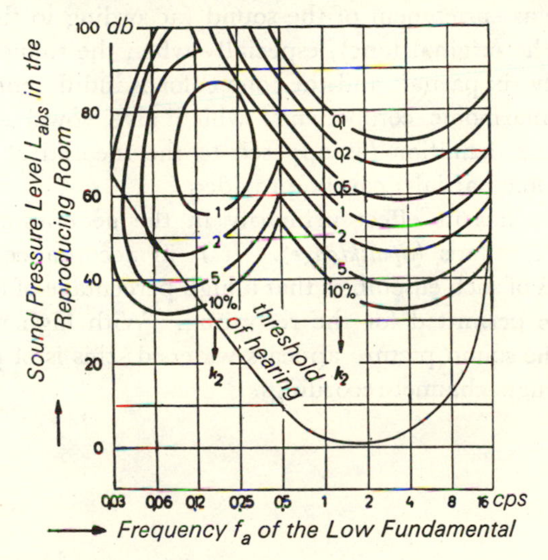

FIG. 94. Just audible distortions (threshold of non-linear distortion) of a fifth, consisting of 10 partials in each case (quadratic distortions) (according to Zwicker).

* Non-linear distortions k_2 and k_3 are caused by overloading of tubes and other electronic components of the transmission equipment.

other hand, with a tone containing 10 overtones, a threshold of 30 db must be passed for the distortion to be noticed. With greater loudness levels, the ear is considerably more sensitive to non-linear distortion; for example, the factor $k_2 = 0.2$ per cent of a c^3 sine tone combined with a fifth is just barely audible, while a tone of the same pitch and with the same distortion but rich in overtones is only audible at 75 db. Further, the curves show that the ear is considerably less sensitive to non-linear distortion of low tones. On the other hand the overtones have only slight intensity in high frequency ranges, so that in practice the ear is less sensitive with speech and music where non-stationary series occur. Experiments with speech have demonstrated that complete comprehension is retained even with a 30 per cent factor of non-linear distortion, but for music the limit is close to 3 per cent, beyond which the tone color is noticeably changed.

As we emphasized in earlier chapters, the absolute removal of every distortion from the musical structure is not at all aesthetically desirable. Small non-linear distortions below 1 per cent can even be regarded as enrichment of the sound (according to the level of strength of the original tone), especially when the sound complex contains only six partials and the tone color is indifferent and soft. Here the inharmonic components, which give vowel colors, for example, their definition (as opposed to the theoretical spectrum of the harmonic partials) can be provided.

Jazz music has its effect primarily in the generation of non-homogeneous sounds (*Spaltklänge*). It is in accordance with the requirements of such ensembles that a high percentage of non-linear distortion be permitted for the recording. With high non-linear distortion, the sound picture appears widened; this is of particular interest in single-channel recordings.

Chapter IX

ELECTROACOUSTIC SOUND STRUCTURE

1. Studio Technique

The majority of the observations presented here are based on the author's work in problems of studio technique and electronic music. The distortion of natural sounds through electroacoustical transmission has stimulated research into the nature of sound. The individual stages in the development of radio since 1923 in the extension of frequency response and dynamics with a simultaneous decrease in distortion have shown the frequency areas that participate in the formation of color in sound, and finally the rôle of noise in music, in the aesthetic sense.

At first it was the deficiency in very low and very high tones that made the sound flat and without substance. Then gradually lows were added to a certain extent, lending a certain fullness to the sound, giving it warmth and contributing to the distinctness of melodic lines in the bass. The sound spectrum was then extended in the highs, adding a certain sparkle which had been missing until then and giving a sharper profile (high fidelity) to the basic color. On this basis an "art of the studio" developed and greater demands were made on artists with respect to the formation of sound than in the first decade of radio. On the other hand, a studio technique developed; as in photography, it became possible to blend in or out lows or highs, sensitively regulating the relationship of direct to indirect sound and thus generating a quasi-stereophonic effect in the formation of shadows and half-shadows in the acoustical sense. In just such a problem complex lies the development of a new "art of tone," a color-register technique, which is not a surrogate for natural sound, but is a new genre with its own attractions—the use of which, however, can easily be exaggerated.

The present development of radio transmission raises some problems which can be regarded as important contributions to the function of sound formation. They involve the expansion of the frequency area of musical overtones up to 15,000 cps. Thus the frequency spectrum of all musical sounds and noises is considered. Earlier, according to international agreements, the low frequency band width of AM was limited to 4500 cps, so that neighboring broadcasting stations would not be disturbed. In Europe this agreement was usually tacitly broken by most broadcasters, who emitted a low frequency modulation of 6000 to 7000 cps. But even so the mutilation of the natural sounds was still considerable, especially in the case of the violin, clarinet, saxophone and organ in accordance with their overtone structure. In the FM radio frequency area there are broader frequency bands available, since there is not such an insufficiency of wavelengths. Through imperfections in the transmission chain (microphone–amplifier–transmitter–receiver–speaker) a loss of the highest frequencies can occur, so that frequently only a frequency area of up to 10,000 to 12,000 cps is actually transmitted.

One notices quite clearly the increase of the highs in FM, for the sounds have more sparkle and are also sharper. The last fact occurs because the higher overtones, beginning with the 7th, are no longer all harmonic (Fig. 1) and piling-up of dissonances increases with increasing frequency. In addition, the opportunity for the formation of intermodulatory distortion is greater with increasing frequency bandwidth. Through non-linearity in the electroacoustical equipment, individual tones and sounds influence one another reciprocally and form new tones, which because of the formation of sum and difference tones may not be harmonic with the original tones. For example, the frequencies 100 and 4000 cps occurring together form new tones of 3900 and 4100 cps. Also, the additional formation of harmonics can greatly alter the sound. When with high loudness level non-linear and intermodulatory distortions of 15 per cent occur and are distributed over ten partials, each receives an increase in amplitude of the order of $\sqrt{0.015} = 12$ per cent, which is noticed as a discoloration of the sound. In the construction of musical instruments it is not as important to have as many high overtones present as possible as it is to have a balanced spectrum, that is, one in which the overtones have the right intensity relationship to one another. One characteristic of very good violins is that the spectrum up to about the 12th overtone is enriched with energy, whereas bad violins, which sound too sharp, have still higher partials of considerable strength.

The enlarged frequency bandwidth results in sharper attacks, because the onset times become shorter in proportion to the growth of the bandwidth.* One can observe further that the areas of high and low frequencies sometimes do not blend sufficiently with one another. This is because one cannot grasp the entire tone frequency area with a single loudspeaker and therefore separate loudspeakers for the lows and the highs must be employed.

It is significant here that low tones are radiated equally in every direction—like the waves from a radio transmitter—while high tones are directed from the loudspeaker within a smaller area. This results in an unequal distribution of energy according to the location of the listener with respect to the loudspeaker. To try to remedy the situation several high frequency systems are placed in different directions, or a larger number of such emitters are arranged around the circumference of a sphere. It is now taken into consideration that in a concert hall it is not possible for the whole frequency area of the musical instruments to reach the listener in full strength. Measurements have indicated that above 2000 cps the maximum loudness which occurs decreases by 6 db per octave (95). It is therefore thoroughly justified to permit a similar curve of frequency decrease for loudspeaker systems, thus considerably reducing non-linear distortion.

All this leads to the fact that many listeners blend down the upper overtones in ultra high frequency radio reception, adjusting the tone control to "bass" just as before. In addition, very high tones after a time are hard on the nerves and result in fatigue (a shrill soprano voice produces the same effect).

It is astounding that not only has the sound picture of musical instruments become livelier in a more extended frequency area, but also that of the human voice, although the enunciation of vowels seldom requires more than 4000 cps. Only the sibilants cover an area up to 12,000 cps. At any rate the extension to 15,000 cps

* According to the relation: onset time $\tau = 1/\Delta f$, where Δf is the bandwidth.

appears superfluous here. The basis for a possible improvement lies in the quite different mode of operation of FM radio. Instead of an amplitude modulation one uses frequency modulation, which has the advantage of reducing considerably the basic noise which exists in every broadcast. Until now very fine nuances in *pianissimo*, and still more the atmospheric noise of the room in which the recording takes place, have been masked. However, it is just these very quiet stirrings that lend personality, expression and localization to the event, and these can now be transmitted, accounting for the fact that the sound picture is now livelier. Modulated quiet noise can have a significant function in music. We do not want to miss the breathing of the singer, Caruso's moaning at the end of a phrase, the light noises of the violin bow, and so forth.

The aesthetic investigation of these relationships is also important for economic reasons, since the widening of frequency bands is expensive. It is therefore desirable to know how far the bandwidth can be limited without causing a noticeable reduction in the quality of the sound reproduction. Experiments in this direction have yielded remarkable results. Figure 95 shows the frequency

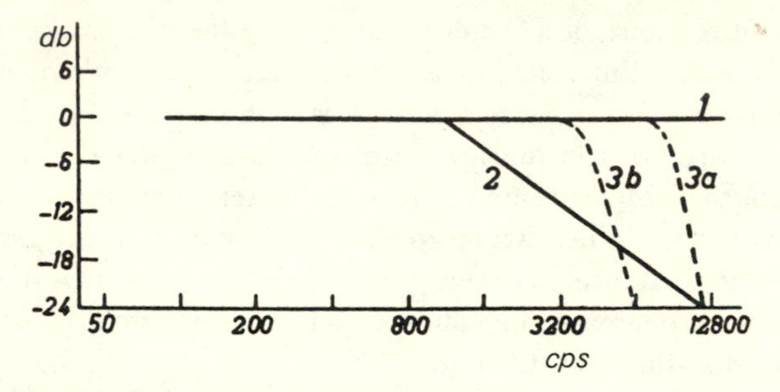

FIG. 95. Effect on the sound for different frequency transmission characteristics (according to Slot).

characteristics of receivers, where 1 represents a linear frequency response up to 15,000 cps, which corresponds to the expectations of high fidelity up to now. Curve 2 represents the principle of the gradual decrease of the highs, as explained above. This is the case for distant seats in concert halls where the sound makes an agreeable overall impression. Sharp cutting to limit the frequency band, according to curve 3, leads to the surprising impression that more overtones are audible than ought to be possible after the frequency cutting. The sound appears brighter than the spectrum would indicate, but at the same time it is harder. Listening to such reproduction for a long time leads to fatigue sooner than in the normal case. Concrete explanations of this cannot yet be offered. The nervous system seems to be more strongly affected for the response curve 3a than 3b (cf. Fig. 95). The latter curve is preferred for the reproduction of old records to reduce the distortions and noise in the upper frequency band.

A further problem of optimal sound formation is presented by the correct placing of the microphone and the complete control of the acoustics in the concert hall. The question is whether the correct sound perspective of a large orchestra is achieved with one or with several microphones. This depends on the even distribution of the sound density in the hall. Certain experiments attempt to arrive at the choral effect (the effect generated by several instruments of one group, e.g., the first violins) through the use of electroacoustical means (for instance, with a reverberation microphone) at a greater distance from the sound source (orchestra) and by this means to allow a reduction of the number of violins. A new phase in the development has begun, so that now not only electroacousticians but also leading musicians are becoming interested in these problems.

2. Electronic Music

> *Ein neuer Klang ist ein unwillkürlich gefundenes Symbol, das den neuen Menschen ankündigt, der sich da ausspricht* (*A new sound is an involuntarily discovered symbol which heralds the new man, who expresses himself therein*).——Arnold Schönberg, *Harmonielehre*, 1922.

We gain a more precise view of the structure of sounds when we work with electronic music, which attempts, on the one hand, to imitate the sounds of known instruments (electronic organ, for example) and on the other hand, to create new sounds, either by the re-forming of random sounds of nature or man-made noises (*musique concrète*) or by a synthesis of elementary sounds, i.e., sine waves, into new tone color scales (see Chapter I). It was seen in the beginning

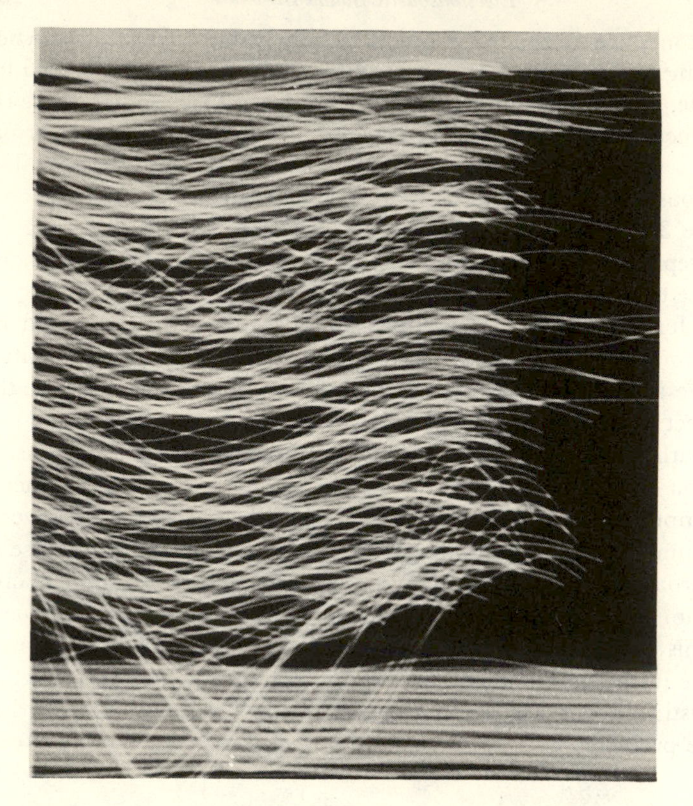

FIG. 96. Oscillogram of an electronic sound with separated periods following each other from below to above. Taken from an electronically distorted trombone glissando
(according to Kusch and Feucht, Technische Universität Berlin).

that electronically produced sounds simply did not correspond to the sound conception of the composers, and it was further seen that the continuous transitions between related tone colors (e.g., viola: cello) were not basically realizable, therefore that the generation of a complete and smooth tone color continuum (p. 85) would not be possible. It was seen how faulty were our theoretical concepts with respect to sound structure and the time behavior of sound.

The main difficulty lay in the electronic generation of a sufficient number of onset transients to produce a sufficient number of tone colors. This lack makes compositions from sine tones tiresome. H. F. Olson was able to imitate with remarkable success the sounds of the piano and wind instruments by imposing the attack and decay curves (Fig. 97) on stationary vibratory processes of variable

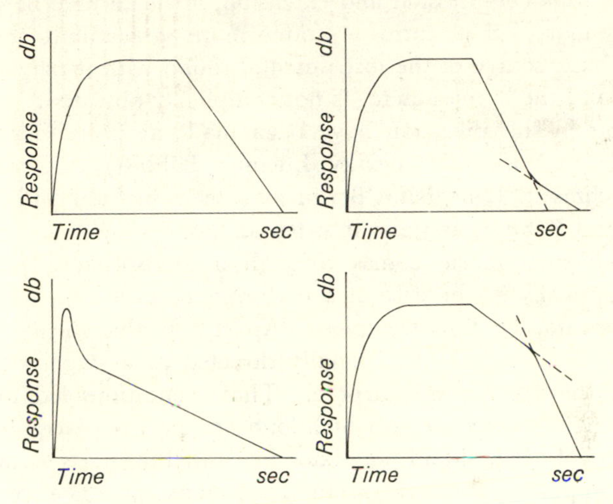

FIG. 97. Growth and decay patterns obtainable with an electronic synthesizer
(according to H. F. Olson).

spectra (74). String instrument colors have not yet been perfectly obtained.

Electronic music has realized means to reverse sound complexes chronologically and also spectrally, and even to fold the spectrum, to compress or expand the sound in time or in spectrum, to transpose it or to move it linearly along the frequency scale (Fig. 75), to split it up by impulses, and finally to add it to other sounds and even to multiply it (interpenetration of various sound events to form a new complex). A new world of sound phenomena has thus been created which at the same time advances our knowledge of the formation of sound in traditional musical instruments. This book has dealt repeatedly with this complex of problems.

Sounds generated by these means are often not sufficiently adapted to the human power of perception, and thus are not understood by the wide public.

Another direction in composition starts out with known musical instruments and attempts through electronic distortion to find new nuances, resulting in a stimulating game of an increasing distance from the instrumental color. The means employed include transposition, spectral modulation and shift, high and low band pass filter, time compression and extension, artificial reverberation and many more. Such forms are often more agreeable to the ear since the basic source of the instrumental sound with its rhythm is retained, and the human factor is not completely obviated.

Attempts in this direction have been made at Columbia University in New York (Ussachevsky, Luening, Babbitt); in Germany at the Technische Universität Berlin by a team including Blacher, Winckel and Krause; and also in other studios.

Since all such music comes through a loudspeaker, nothing happens optically in the case of a performance in a hall, and the listener does not feel fully occupied. Apart from this, the arrangement of seats in a hall traditionally directed at a stage without events for the eye has no purpose. These conditions led to the development of *space music*, in which loudspeakers are placed in the four corners of the hall and play back four different parts arranged by the composer in a contrapuntal relationship—or a contralocational relationship. It is surprising to what extent the sound texture is rendered more relaxed and, because of the space dimension, made more acceptable to the audience. This principle is also employed in the dispersion of several orchestras in one hall, which was popular in the Baroque period and was later employed by Berlioz in his *Requiem* and today by Boulez, Stockhausen, the composers of the Berlin school and others.

To discuss in detail the problems of electronically generated sounds would exceed the scope of this book. A detailed literature exists on this subject (65, 115, 118–121).

Chapter X

THE EFFECT OF MUSIC ON THE LISTENER

Music is sounding form in motion.
E. Hanslick (1825–1904).

Summary

The investigations discussed in this book have shown that not only music as a whole but also its individual component—sound—is an element exclusively in motion. If one were to permit a single sound to be intoned undisturbed and unmodulated for a certain time and without the addition of other sounds, it would soon be "forgotten" by the psyche and therefore would no longer exist for our consciousness. A continual monotonous hum of a machine in a factory disappears from the consciousness and is noticed again only when it is turned off. The ticking of a clock drops quickly from our consciousness, and will be clearly noticed only at the moment when it stops, because that moment represents a discontinuous transition between two states. A long-held organ point has no independent function; only in conjunction with the upper voices can it accomplish a harmonic modulation. Since in earlier centuries it was not yet customary to enliven a long, steady tone by means of *crescendo*, trills and vibrato, the technique of ornamentation and coloration was extensively employed. Evenness of beat would also lead to rigidity of music. It is the fluctuation in the meter, the agogic molding, which "disturb" the frozen pattern of synchronization and breathe life into the work of art. It is this which creates the inner motion, makes contact with the listener and produces the visible or audible event.

If we examine sounds that occur in nature, we look in vain for sounds which remain even for a short time at the same pitch or are constantly of the same color. For the communication of all beings such sounds would indeed be meaningless, for they contain no information, only a repetition of the first impulse. A signal forms its characteristics even when only briefly intoned, and even when it is a call for help it is more noticeable when it is interrupted many times (cf. the wobbly tone of the air raid warning siren in Europe as opposed to the constant sound of the all-clear siren, the relief). Out of the varied world of animals' voices we select here the song of birds, because it is considered especially musical. If we observe the sounds on oscillograms (Figs. 98 and 99), we see that there

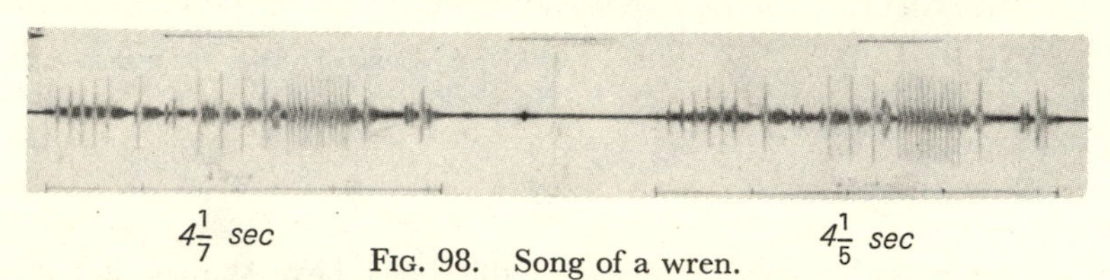

FIG. 98. Song of a wren.
Demonstration of the rhythm by the sound pressure recording.

is not simply an uninterrupted change in sound formation forming a melody, but, surprisingly enough, the melody is in no way random or without rules, but rather is repeated in fractions of seconds exactly note for note in the same rhythm, as will be seen in the example of the wren. It is also to be noted that a small child in his first years of speaking does not utter lengthy constant sounds, and, as a matter of fact, is unable to do so. The special technique of the physiological functions required here (breath control, training

FIG. 99. Song of a nighthawk.
Pitch recording on the sonagraph.

of the muscles of the larynx and of the nervous system) is only acquired after long years of practice. Human *speech* is according to its structure also an uninterrupted series of transients, so that it is not justifiable to speak of speech-"sound" (*Klang*), as unfortunately is still done in the field of phonetics. Normal conversational speech contains on the average four or five syllables/sec, which continually flow one into another and even in these transitions contain information.

How are we going to define *singing* as opposed to speaking in terms of our investigations? Singing is the development of utterances of speech into a cultivated sound through the extension of the vowels in time, mostly on a higher pitch level. The vowels can predominate over the consonants to the point where one has a pure *vocalise.* According to the style and character of the vocal music, all intermediate stages from *parlando* and recitative to aria and chorale occur. The more song develops in the direction of vocalization the lower is the information content of a composition. The expression content must therefore be increased in equal measure, which is achieved, for example, by larger intervallic leaps in the melody. This means greater unsteadiness in the transitions and more abrupt changes in the color registers of sound, psychologically an increase of tension. In addition, we have a more intensive modulation of the sustained vowel intonation, rhythmic changes and so forth. How difficult it is to satisfy the ear by sufficient density of events may be seen in a quotation from Helmholtz: "Monophonic song does not appeal to us any more, it seems empty and incomplete. On the other hand, we are satisfied even if only the twanging of a guitar adds the fundamental chords of the key and indicates the harmonic relationships of the tones."

How our aesthetic feeling prescribes the resolution of a "straight tone," of a strict periodic series, into an elastic configuration—a more plastic remolding of the rigid framework of tones arranged according to the rules of harmony and counterpoint—this can be seen objectively in the photography of music or speech through apparatus that register tone series, e.g., the oscillograph, level recorder and sonagraph. The latter, which is also known as the Visible Speech process, shows the time flow of a sound phenomenon as a transient. The frequency scale—in the limited area of 85 to 12,000 cps—is entered on the verticals, the spectral lines of varying lengths being replaced by dots of varying blackness (see Figs. 38 and 50).

On a piece of paper stretched over a rotating drum head, lines for frequency areas with a width of 300 cps are inscribed for the time flow of speech or music of 2.4 sec duration with an electronically activated writing tool. Also necessary is an electric filter component which switches, from line to line, a further 300 cps band filter (Fig. 100).

The very prominent black bands which occur in a sonagram are identified as especially strong partials in the spectral construction of the sound, that is, as formants (see p. 13). The lowest band, as fundamental, shows the movement of the melody; the second and third are of particular significance as the color-determining formants. One sees that there are no straight lines, that is, no absolutely

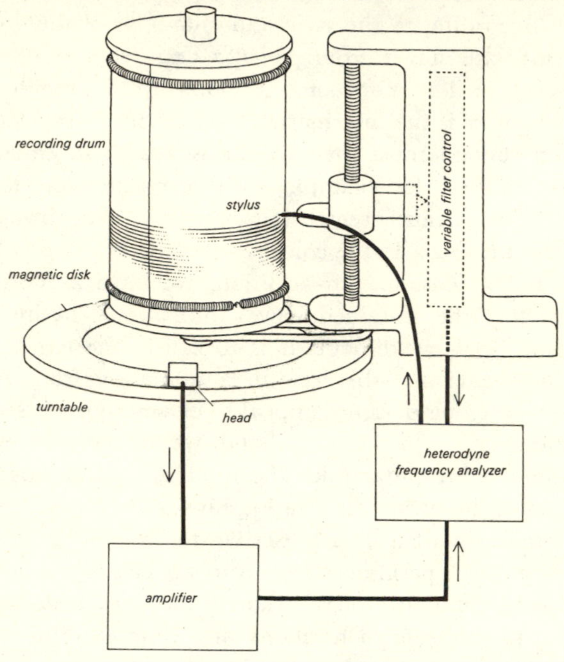

FIG. 100. Sound spectrograph (sonagraph).
A sound of 2.4 sec duration is transferred to the magnetic disk and then repeatedly sampled as the disk rotates with the recording drum. The stylus passes a spark to the drum, etching the recording paper.

stationary sounds. Uninterrupted changing of speech progression and music flow is characteristic, obtaining even in the smallest time intervals, where the actual forming of the substance for purposes of information occurs.

From studying the Visible Speech diagrams, we see how the transients achieve full aesthetic effect at a critical optimum, but in excess or insufficiency reveal deficiencies in the voice sound.

The movement of sound in music—clearly seen in the sonagram—is physically a time, frequency and amplitude modulation whose limits are known to us today with respect to its musical aesthetic value. These phenomena are shown theoretically in the onset transients and in practice in the attack and further in the frequency modulation of vibrato, tremolo, trills and other figuration, the amplitude modulation in *crescendo* and in the dynamic expression and so forth. Frequently frequency and amplitude modulation are combined. Changes in sound achieved in this way influence the tone color in countless variations, as already established.

In sound emission the written note value never corresponds accurately to an exactly defined vibration frequency, but rather to a "*frequency band*" of vibrations, where the written note simply indicates the average pitch. Our note-head symbolism is far from showing the actual relationships in music. Apart from the fact that the frequency band depends on the resonance width of the sound emitter (musical instrument), one attempts in performance to widen the band through modulation effects, such as vibrato or trills. A certain amount of spectral energy, a lowest-level stimulation quantum, is therefore important for hearing. This cannot be covered by a single sine wave, since for physiological reasons the amplitudes can assume only limited values.

The concept of a frequency bandwidth for the single note value corresponds to the principle of uncertainty better than the concept of a single vibration of a certain pitch does, since a tone cannot be precisely determined physically for frequency and duration simultaneously (see p. 49f.). An attempt to do this would lead to a blocking of the nerves as a result of fatigue.

In addition, the pitch differentiation sense in the ear is not so well developed that it can determine a pitch exactly. Intonation fluctuations within certain bounds are not noticed. In the purely theoretical considerations up till now the demands of pure intonation have been overestimated. The different tuning systems are only slightly different from each other as far as hearing is concerned. *Tempered tuning* proves to be much more stable in practical musical performance than pure intonation, for instance, because unintentional distuning, such as bad intonation, faults in the instruments and so forth, is more disturbing in pure than in tempered tuning, which permits a certain area of fluctuation. The maintenance of this fluctuation area is important for the performance of music, since it creates an inner modulation which supports the sound (combination tones and beats). Of course, it then becomes very critical, since a slight exaggeration is immediately recognized as bad intonation.

From the aesthetic standpoint it might be feared that the distuning would lead to sound dimming, to impurity of sound, to noise. However, further researches have shown how important the dimming and the noise are. In painting, too, we know the principle of covering colors, making them intentionally "impure" by mixing in order to provide differentiated tones within one basic color, even to mask the entire picture behind a veil so that it appears as if seen through smoke, which is what Leonardo da Vinci referred to with the term *sfumato*. The painter's "complex sound" thus becomes richer in information.

In music, too, intonation and color formation obey similar aesthetic laws of deviation from strict harmony and perfect form. It is becoming more clear to us now that noise is as important as are consonants between vowels. It is physically unavoidable, but its use in practical music extends beyond the minimum. Noises are present in the continual transients; they can profile or smear the sound picture, and thereby give music the power of communication.

Thus we come closer to the question: What actually is the communicative power of music? What is its effect on the listener? In a way, music is a double language. First, there is the series of tones, irrespective of the formation of intervals or melody, rather like series of syllables in speech, likewise disregarding speech melody. This experiment was made with the help of the synthetic speech generator Vocoder (monotonous speech). Then, there is something quite different, namely the tone simply a fixed signal, without further information content of other parameters, and the tone relationship conceived mathematically, for instance in the Pythagorean sense. That is the other language, controlled by *logos*, but not the subject of this book. Thus there are a Pythagorean and a poetical principle alongside one another (W. Gurlitt); in the course of history, each has from time to time been dominant. This is also related to the development of musical instruments; for the tones of a dulcimer and of a guitar do not have the same rich and variable structural content as, for example, syllables of speech. With increasing cultivation of instrumental sound and, the collaboration of instruments in the orchestra, the sound language has gained the upper hand.

Eggebrecht points out that from the end of the *ars nova* around 1420 up to the appearance of "atonal" music after 1900, the dominance of the tone as signal declines in favor of color of the tone, finally becoming completely lost (17).

There is still another point of view to discuss, which pertains to the nature of music. We have already seen that the feeling of formal order which we receive in listening to music does not stem from an all-embracing harmony of the artistic work in a strict mathematical sense. On the contrary, because of the stationary character of the highest principle of order, which is rooted in a contrast between the proportions of small integers, it leads to the concept of infinity, a harmony of the spheres, which the senses can no longer grasp and—if an approximation were possible—would then exclude our participation.

At the root of the phenomenon of harmony lies the strict periodicity of every progression. It is precisely this which must be avoided in music, as experience shows. Thus we have seen that the quite elementary entity, the sine wave, does not exist for us (p. 24f.), and that the pure intervals of the triad of simple tones do not evoke a musical experience, but on the contrary actually require a stimulating component—at least the 7th partial—in order for a vital and satisfying triad to be formed.

Thus we come even closer to the goal of the harmonic ideal, but we can never attain it since it would then elude our consciousness. We have seen further that the strictly periodic partition of time leads to rigidity of sound impression, that strictly periodic vibrato and the corresponding tremolo and trill coming from a tone generator are unbearable, and even that the series of reflections of sound from the walls of rooms which combine to form reverberation must not be periodic, therefore that even the architecture of the room in which music is to be heard must be constructed in accordance with acoustical and musical principles. Experiments with synthesized music have established the truth of this. Periodic organization would impose a rigid law on a work of art from outside which would make human creative power illusory or would be prejudicial to its operation.

When a musical revelation is called "divine," a very human god is meant, one who speaks to us in the idiom of fluctuating human nature, for only in terms of these same sounds, related to us, can the soul be reached by means of the senses. The "harmony of infinity" will never reach our senses, and only similes can give us an idea of it.

Of what type are the fluctuations, that is, the deviations of periodicity, with which we have become acquainted in the diagrams, registration strips, mathematical and physical deductions? We have indeed established an admissible fluctuation breadth, but we have not as yet found laws.

We find a hint in an electroacoustical experiment. If one cuts the amplitude of the vibration curves of a spoken sound in a process which turns it into a rectangular curve, the understandability is thereby scarcely impaired. Accordingly, the distances cut between the zero line and the sides of the vibration curve are important for the information content (Figs. 101 and 102). The case is similar

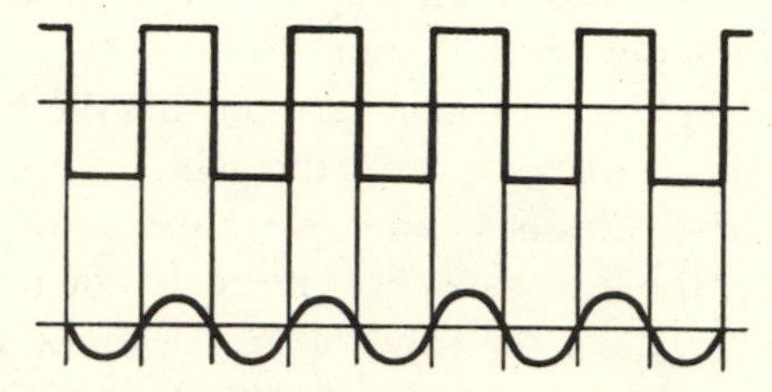

FIG. 101. Repeated peak clipping and amplification of a sine wave produce a rectangular wave.
The original zero crossings of the horizontal axis are preserved.

when the original curve (*a*) is subjected to a process of mathematical differentiation (*c*), the peak clipping process according to Licklider, referred to by Laey and Saxe (49).

The information transmitted by such vibrations—regardless whether it is semantic or aesthetic in nature—is given not through periodic but through irregularly distributed zero crossings of the curves (Fig. 102). In this rhythm, impulses are transmitted through the nerves to our perception organs. Stress and accent, or the emotional content transmitted by them, are given in the non-periodic time division.

The perception of small fluctuations in a determined series, as the characteristic phenomenon of semantic development, agrees with the basic tendencies of our nervous system and its time constants. A diagram of the impulse transmission in the network of the nerve fibres is given, for instance, with the help of Figure 64, but it must be recognized that we are concerned here with a very simplified presentation of a basic framework of relationships, and moreover the finer modulatory behavior could not be taken into consideration since it still has been too little investigated.

In any case, we know that impulses are not only sent forth through electrochemical transformation, connected with the nerve fibres, but also exist in the form of electrical fields, which go beyond the limits of the individual neurons and influence their excitability positively or negatively. Thus a second level of nervous activity exists, which exerts an influence on the numerous cross contacts of the fibres running toward the center. From this is derived a

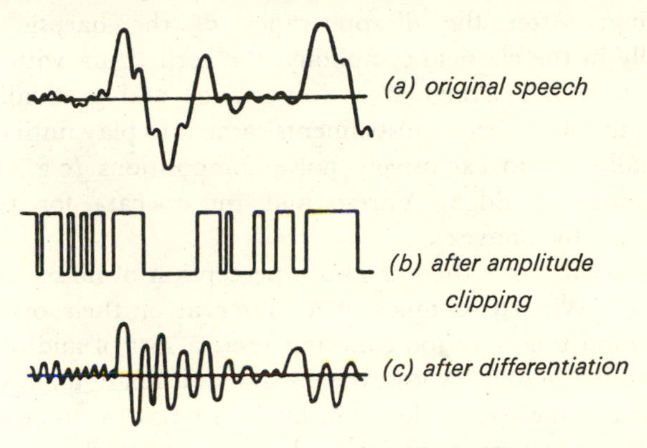

FIG. 102. Repeated clipping (*b*) and differentiation (*c*) of speech curves (*a*).

fine scale of excitability states, which is further influenced by the hormone regulation of the synapses in the transmission network of the nerves. Thus man's whole state of being is interrelated, through the circulation of the blood and the hormone system (analogue system), with the signaling system (digital) leading to perception. Accordingly, the impulse series, as an image of the acoustical environment, is modified by fluctuations which are caused by other environmental influences and individual conditions.

There are tiny deviations from periodicity which have significant effects, as mentioned before. They are subject to a statistical behavior and thus express the multitude of individual possibilities. They elude exact classical mathematical analysis. More success is obtained with statistical mathematics. Nevertheless, we can construct only a formal scheme, a framework for our work of art, but sensations are not fully explained by causal connections. "Cette irrégularité donne précisément cette richesse," says Pablo Casals. "Tout est irrégulier et différent, mais combien magnifique est l'harmonie de toutes ces choses."

It has been proved physicomathematically that every attack of sound is accompanied by noise, and the more suddenly it appears the more noise is present. A composer unconsciously makes use of this; in the Baroque period, for instance, the hard attack of the harpsichord strings supplied energy to the cantilena being accompanied; in other words, a function of disturbance leads to an enlivening. After the disappearance of the harpsichord and especially in the classical symphony, the kettledrum with its broad noise spectrum assumed more importance and gradually an increasing number of noise instruments came into play until exaggeration finally led to exclusively noise compositions (e.g., the noise compositions of Edgar Varèse and the toccata for percussion instruments by Chávez).

In a certain sense there is also a perception of noise involved in *dissonance.* When two tones of an interval or their overtones or combination tones are too close together, a sort of audible friction effect results; in other words, the conglomeration of tones which are too close together (tone cluster in modern music) is to be viewed as an energy enrichment, and thereby as a spectral concentration. Besides, very low lying difference tones, because of the significance of the low frequency transients, attract the attention of the listener.

Observed aesthetically, a musical work (of any period!) represents a thought-out distribution of "good sound" and noise—an analogue to vowels and consonants—in which the statistical distribution may fall either in favor of the "good sound" or of the noise. Accordingly, one can consider the musical work more as a sound painting or more as speech (message). The differentiation of the acoustical complex in the one or the other direction occurs psychically in the sphere of hearing. Ontogenetically, the sound perception appears to be primary; the apprehension of melody as a collective concept for pitch differentiation, linear development and, further, perception of form are secondary.

If one investigates the onset transients of different sounds in general, one can establish that, seen mathematically, they represent a natural phenomenon in any case but that this process can be extended by the mechanical onset time of the sound generator as well as by the onset behavior of the room in which the sound is intoned—and finally that it can be modified further by the characteristic attack of the player as well as by the attack or operation noise of the instrument. In addition, the ear's perception must be taken into consideration; it, too, "attacks" with a certain time constant (Fig. 72). But if one ignores the fact that the onset is theoretically infinite and lasts quite long even in practice (at least one second if nothing but the onset reverberation of the room is considered), then one can limit the significant changes of the sound complex during the formation of the sound to $\frac{1}{10}$ sec.

With a view to these observations, one finds in inspecting musical literature that very frequently two tones occur at least $\frac{1}{10}$ sec apart, for instance in Bach's fifth *Brandenburg Concerto* (Fig. 103), where the

Fig. 103. J. S. Bach, *Brandenburg Concerto* No. 5.

thirty-second notes occur $\frac{8}{100}$ sec, that is, barely $\frac{1}{10}$ sec apart. If the simple notes of a melody are arranged in a slow tempo, it is necessary to add one or more parts *punctus contra punctum* to get an energetic saturation in the nervous system, in this way arousing attention for the perception of the music. Notes of longer duration have to be contrasted by florid effects in other parts. In Occidental music the added parts to a *cantus firmus* have the functions of linear counterpoint, influencing the harmonic progressions or emphasizing

Fig. 104. C. P. E. Bach, *Allegro*.

the rhythm by accents. A distinct measure for richness in music patterns is given by information theory. If this requirement of a desired disturbance function is not satisfied, the held tone usually will be modulated dynamically or by vibrato; the performer follows his feelings in this more or less unconsciously. If even this is not the case, as in Figure 105, the sustained unison sound serves to increase the contrast of the following abruptly cut-off dissonant chord; the long rest is required for the process of psychic assimilation. During this time the dissonant sound complex is stored up

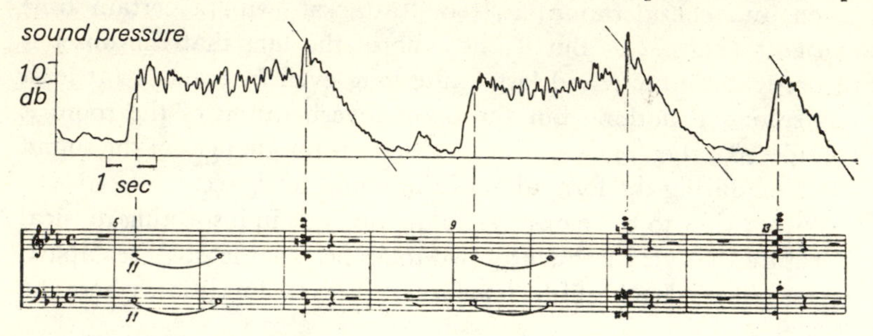

FIG. 105. Beethoven, *Coriolan Overture.*
Above: the reverberation process.

as it decays by the reverberation of the room. There is no such thing as an absolute silence in music. A composer should be aware of this when he prescribes a *fermata*. We see through more exact observance of the level diagrams of sustained tones that this is not a strictly periodic event. The loudness fluctuates by more than 10 db, as Figure 105 shows.

The significance of the rest is greater in non-tonal music, because the expressiveness with which every sound combination is charged requires a subsequent interval for psychical assimilation. Figure 106 presents a complete piece of music with a high information content, insofar as its performance is at all possible considering the

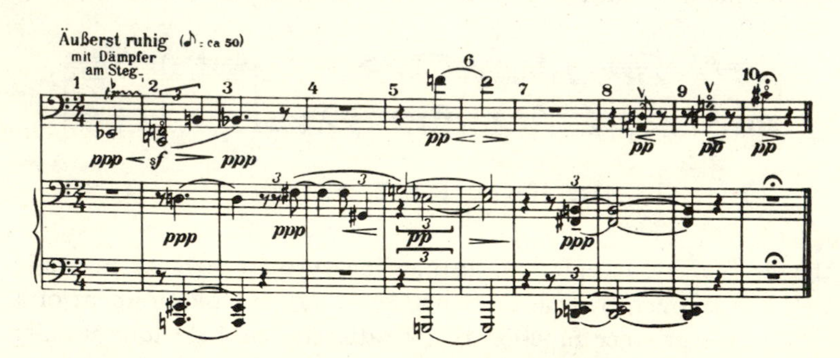

FIG. 106. Webern, the third piece from *Drei kleine Stücke*, Op. 11, for cello and piano (Universal Edition).

low sound level called for and the basic background noise of the room. The range covered is only the bass range from 41 to 349 cps—a range in which distinctness of pitch is reduced (see Fig. 76).

Another musical example (Fig. 107) shows how a certain fundamental noise with dominant character below the strings—tympani *pianissimo* for 35 measures—influences the atmosphere of the sound space. It is much less than a whisper or forest murmur, as in Richard Wagner's *Siegfried*, for instance. The *pianissimo* of the strings, as the usable signal, is scarcely heard above the noise background, as the communications engineer would say; the engineer,

FIG. 107. Bruckner, Fourth Symphony in E-flat Major.

conversely, attempts to lower the noise so much that a high usable signal is obtained. We must remember that we can never achieve absolute quiet, for even "holding one's breath" at suspenseful moments does not exclude the accidental presence of some room noise. In concert halls a value of about 40 phons is to be given for this.

A further example shows how the entire sound mass of an orchestra with choruses can be crammed together into a "gray" blending for the listener, often unconsciously. This can be occasionally heard in Bach's B Minor Mass, and can even be clearly proved if one cuts short test pieces out of a magnetic tape recording. It is grandiose how Bach then employs a bright, gleaming trumpet over these masses of sound.

In this connection the question may occur how one differentiates the individual parts of an orchestra with the ear if they are distributed over a wide frequency area and, together with all their overtones, produce a complete spectral filling-out of the frequency scale which would have to be perceived as a noise. If all the instruments were to play constant tones of different pitches but of one certain duration, this effect of gray blending would be generated. But since every instrument has its own characteristic onset transients and, so to say, lives its private life for the duration of its sounding, the inner ear can, through analysis, already identify it, unless fatigue sets in because of the sounding of too many instruments (masking, see p. 145; blending, see p. 71).

Let us keep in mind that the stationary sound value, as it would occur if it were reproduced as it is in the score, practically never reaches the ear, for music is *constantly in motion.* The transients described in this book provide another tone color than that of a stationary value. Anyone can test this out by playing a rapid piece with 64th notes on a string or wind instrument first in a normal tempo and then extremely slowly, so that every note is held. The tone color is changed slightly. Thus it is not the tempo or the changed rhythm alone which causes the different impression. One can say, conversely, that the changes in tempo give the transients more or less opportunity to have an effect and therefore change the tone color. Thus the rhythm at any given moment indirectly produces the nuances of the tone color!

The basic difference in the techniques of two conductors may lie in such relationships (see p. 47). Sharp attacks and quick tempi make music noisier. The layman appropriately calls such a performance "dizzying" or "intoxicating."* The composer who is able to hear such differences contrasts very active passages with very sustained phrases (Fig. 108). Even the piano can be used for

FIG. 108. Beethoven, Sonata, Op. 10, No. 2.

* [The German adjective used here (*berauschend*) is linguistically related to the German word for noise (*Geräusch*).]

register differentiation, and this can be achieved in the composition itself, i.e., apart from the pianist's use of touch. On the other hand, the pianist can consciously or unconsciously influence the sound picture, for instance through a delayed attack by the right hand as compared with the left. The attack as a whole is thereby lengthened, and this may have a disturbing effect on the listener. These disjoint attacks are used in a positive sense by the composer in arpeggio passages and also in simple grace notes, especially in passages where this forming element is in the foreground as opposed to the sound element, i.e., where the communicative power is to be greater. This is particularly clear in the case of song, in which the music should have an immediate speaking effect through the words. An agitated, expressive statement does not begin soberly and precisely at a prescribed pitch, but is an outcry, in which the pitch glides upward, or a falling-off of tension, a groan, in which the pitch glides downward.* The human voice expresses all states of excitation with gliding series of tones, that is, with a considerable gradient of frequency change, which is related to the sudden muscular tension of the glottis. For want of a more suitable notation, Franz Schubert marked such expressions of feeling in his songs with grace notes, which in this case glide into the main note (Fig. 109). The

FIG. 109. Schubert, *Winterreise* ("Der greise Kopf").

Trautonium (the electronic instrument invented by Trautwein) reproduces such excited expressions with gripping reality (see, for example, its outcries in Honegger's *Jeanne d'Arc au bûcher*). From this viewpoint it is understandable why a feature like Caruso's typical moan at the end of a phrase is perceived as exciting, although it is a noise and against the rules of vocal training, or why a vocal

* [The prefixes *auf-* and *ab-* of the German words *Aufschrei* and *Abspannen* here translated as "outcry" and "falling-off of tension" convey the notion of a movement upward and downward, respectively.]

smearing, of course within bounds, is sometimes felt to be not unpleasant. We could list here numerous examples from the performance of music, but the reader is now able himself to point out interesting cases. The essential point is not to be prejudiced by the written page of music and to rely entirely on the *ear as the sole authority and court of appeal.* Then, by a growing refinement in one's listening ability, one will constantly make new discoveries which can be explained very naturally without entering into complicated explanations or metaphysical interpretations. Music, too, is down to earth: all sounds are generated by concrete vibration forms the laws of which are quite well known today.

The increasing exactitude of musical performance has been brought about through the microphone. It is the sensitivity of this "electronic ear" and thus the preference for the primary sound with considerable suppression of the secondary influences, namely the room reverberation, which has the effect of a beneficent adjustment of unclean performance. This more accurate listening by means of electronic amplification has doubtless been a good education for better achievements in the area of music, but in its development toward absolute exactitude—toward the ideal form—a work of sound of nearly mathematical abstraction has come into being, for we miss something of man's far from perfect behavior, the fluctuations in his mind, in other words his deficiencies of various kind. The extent to which the "fluctuation width" is necessary for the hearing process should have been established in this book. In this sense we can also understand why the ideal form of a Greek Venus is often not as gripping as a statue of Käthe Kollwitz, for example, if we ignore absolute artistic evaluation for the moment.

Music which is being heard must be evaluated differently than the abstract notation of the score. Herein lies the failure of so many composers, who do not take the effects conditioned by the listener as their point of departure, but have become involved in the external function relationships of a language of symbols in musical notation. Contemporary scientific investigation and the derived experimentation in the area of art will lead to a healthy clarification and refinement of the various opinions and, it is hoped, will provide the stimulus for new creative accomplishments.

Appendix

THE AUDIBILITY OF ONSET TRANSIENTS

Explanation of Figure 110

We shall attempt here to test the theoretical relationship that we have found with respect to the onset transients, to see to what extent they are practically audible. The sound of a source must pass through a series of links before it reaches the consciousness of the listener. The sound of the musical instrument, which is released through a voluntary act of the brain, spreads out in the room and reaches the ear not only as primary sound directly, but also as secondary sound, as reflections from the wall, and the ear itself then becomes a chain of transmission links (outer, middle and inner ear), in order that the sound may then penetrate into the consciousness in the brain.

Every link of this transmission chain has its own characteristic onset time constant. The resulting audible onset transient complex is found from the sum of the squares of the individual time constants:

$$\tau_{\text{res}} = \sqrt{\tau_1{}^2 + \tau_2{}^2 + \tau_3{}^2 + \cdots}$$

The increase in the sound pressure is represented in Figure 110 as an exponential function. Since the perceptive power of dynamics operates as a logarithmic function within certain limits, the two curves compensate one another, so that finally there is a straight line. The increase in loudness at the onset of a sound is perceived as a nearly linear increase.

We will begin with the onset transients of the musical instrument. Any value from 10 to 300 ms can be assumed according to the type

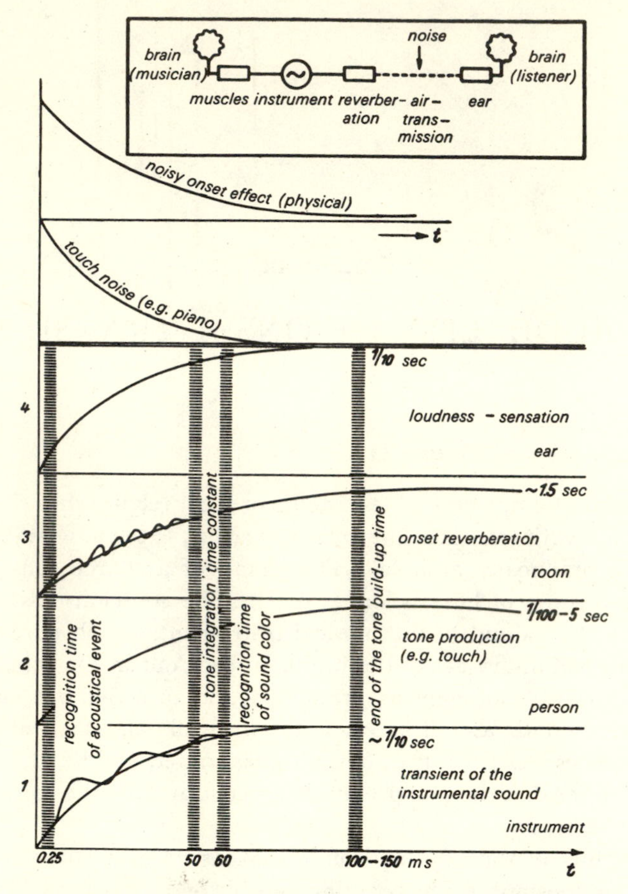

FIG. 110. General view of the internal and external influences on tone sensation in connection with the build-up process of sound in time.

of instrument. We will select as a statistically frequent case an average time of 100 ms $= \frac{1}{10}$ sec. The ideal onset time of the instrument is extended by the individual type of touch or bowing by the player, or the attack and dynamic modulation of the singer. We shall assume for this a time constant of $\frac{1}{10}$ to 5 sec. Further, we must consider the onset and reverberation properties of the room, which for a concert hall may have a value of 1.6–2.0 sec. In the case of greater distance from the sound source the build-up time of the sound is extended by the reflections. Finally one must

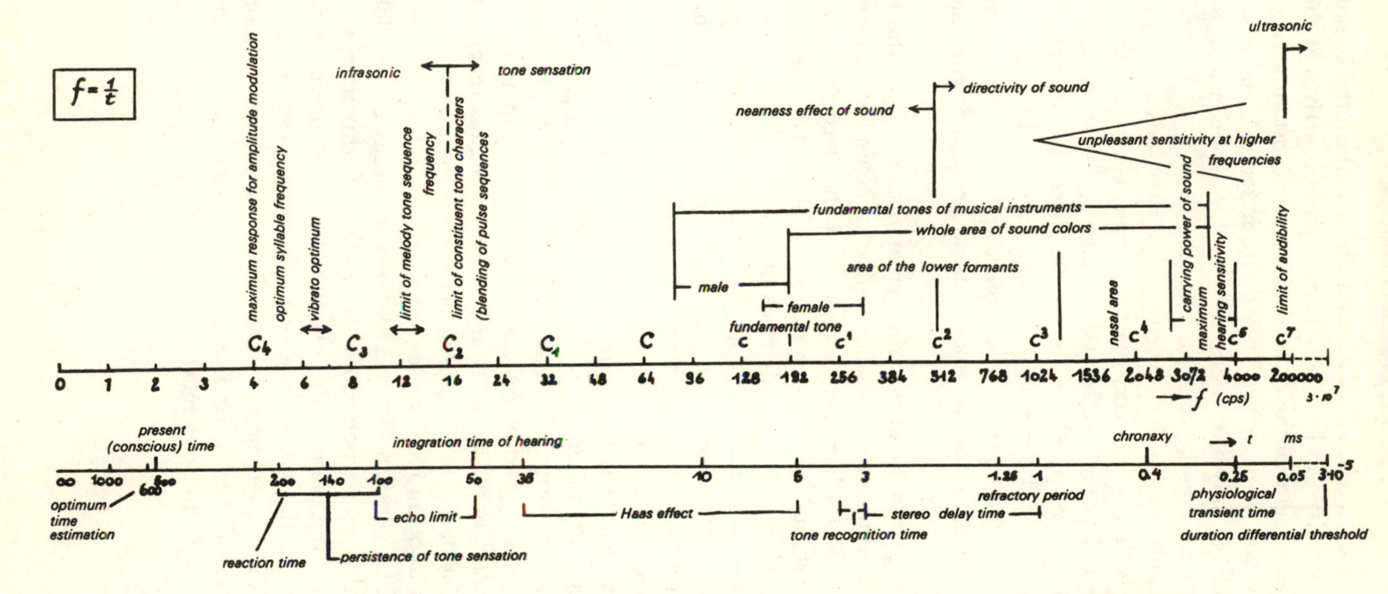

FIG. 111. The psychoacoustical characters in time and frequency scale in rough limits of effectiveness as a summary of the hearing functions discussed in this book.

remember that a sound event that begins abruptly is not immediately perceived in its original strength, as it arrives at the ear, but with the integral of the physiological time constant. A *pp* tone of 20 phons will be perceived after 65 ms in its real loudness, a 90-phon sound (*ff*) only after 140 ms. For this time constant, too, we shall assume an average value of 100 ms.

It is possible that an instrument with a short onset time, e.g., a reed pipe with $\tau = \frac{1}{100}$ sec, may be immediately perceived with nearly the full strength of its sudden attack, if there is no playing delay and if perception occurs in the direct sound field. The mechanism of sound perception itself adds a delaying link of nearly $\frac{1}{10}$ sec; this may be regarded as a protective device, because without it speech as well as music would all be perceived with clicking attacks.

If we happen to be at a rather great distance from the sound source, or even in a sound shadow—that is, behind a pillar or in a similar situation—then the secondary sound has an important influence and must be considered in calculating the time constant. In the case of an instrumental attack of $\frac{1}{100}$ sec and a reverberation time of 1.6 sec, the reverberation time is so much more important that the instrumental attack is suppressed and the instrument then loses much of its individual character. This is in fact the case in large live rooms, such as cathedrals.

A further characteristic magnitude is the tone recognition time of about 3 ms in the minimum which was discussed on p. 111.

Nature has so ordered things that the human organs of sound emission, speaking and singing, have the same damping (e.g., that of the mouth cavity) as the sound-receiving organs (ear). From a damping value of $\delta = 1.3$ one derives an onset time of $\frac{1}{10}$ sec. This seems to be an ideal average value having a favorable effect.

The highest curve in Figure 110 shows the decay of the click-like noise as a physical phenomenon. In addition to this noise, we must also consider the noise of the playing of the instrument, e.g., the touch noise of the piano tone. It also decays exponentially, as shown in the second noise curve. These two noise phenomena mask the sound during its formation and make tone recognition and sound differentiation more difficult.

BIBLIOGRAPHY

I. General Works on Hearing and Perception

(A) Békésy, G. v., *Experiments in Hearing*, McGraw-Hill Book Co., Inc., New York, 1960.

(B) Bergeijk, W. A. van, *Waves and the Ear*, Anchor Books, Doubleday & Co., Inc., New York, 1960.

(C) Broadbent, D. E., *Perception and Communication*, Pergamon Press, London, 1958.

(D) Helmholtz, H. v., *On the Sensations of Tone* (4th ed., 1877), Dover Publications, Inc., New York, 1954 (Eng. trans.).

(E) Jeans, Sir J., *Science and Music*, Cambridge University Press, Cambridge, 1937.

(F) Meyer, M. F., *How We Hear*, Charles T. Brandford Co., Boston, 1950.

(G) Rayleigh, Lord, *The Theory of Sound* (2nd ed., 1894), Dover Publications, Inc., New York, 1945.

(H) Richardson, E. G. and Meyer, E., *Technical Aspects of Sound* (vol. III), American Elsevier Publishing Co., New York, 1962.

(J) Rosenblith, W. A. (ed.), *Sensory Communication*, M.I.T. Press, Cambridge, 1961.

(K) Salzer, F., *Structural Hearing*, Dover Publications, Inc., New York, 1962.

(L) Skudrzyk, E., *Grundlagen der Akustik*, Springer-Verlag, Vienna, 1954.

(M) Stephani, H., *Zur Psychologie des musikalischen Hörens*, Bosse, Regensburg, 1956.

(N) Stevens, S. S. and Davis, H., *Hearing: Its Psychology and Physiology*, John Wiley & Sons, Inc., New York, 1938.

(O) Stevens, S. S., *Handbook of Experimental Psychology*, John Wiley & Sons, Inc., New York, 1951.

II. Current Literature

(1) Adorno, T., "Vers une musique informelle," *Darmstädter Beiträge zur Neuen Musik*, B. Schott's Söhne, Mainz, 1962.

(2) Autrum, H., "Die Zeit als physiologische Grundlage des Fernsehens," *Studium Generale*, 8:527 (1955).

(3) Backhaus, W., "Die Ausgleichsvorgänge in der Akustik," *Zeitschrift für Technische Physik*, 13:31 (1932).

(4) Bechert, K., and Gerthsen, C., *Atomphysik*, Goeschen Collection, W. deGruyter, Berlin, 1944.

(5) Békésy, G. v., "Skin Sensations and Hearing," *Journal of the Acoustical Society of America*, 27:830 (1955); 29:489 (1957); 29:1059 (1957).

(6) Beranek, L., *Music, Acoustics and Architecture*, John Wiley & Sons, Inc., New York, 1962.

(7) Bergson, H., *Durée et Simultanéité*, Paris, 1922.

(8) Boré, G., *Kurztonmessverfahren für Lautsprecherbeschallte Räume*, Technische Hochschule, Dissertation, Aachen, 1956.

(9) Bürck, W., Kotowsky, P. and Lichte, H., *Elektrische Nachrichtentechnik*, 12:326 (1935).

(10) Bürck, W., *Grundlagen der Elektroakustik*, W. Sachon, Mindelheim, 1953.

(11) Bürck, W., "Das Weber-Fechnersche Gesetz in der Akustik," *Radio-Mentor*, 20:560 (1954).

(12) Boulez, P., *An der Grenze des Fruchtlandes*, Die Reihe No. 1, Vienna, Universal Edition, 1955.

(13) Brock, P., "Die Bedeutung von Rhythmen für theoretisch-biologische Zusammenhänge," *Naturwissenschaftliche Rundschau*, 3:349 (1950).

(14) Caine, H. le, "Electronic Music," *Proceedings of the IRE*, 44:457 (1956).

(15) Cremer, L., *Geometrische Raumakustik* (vol. I), S. Hirzel, Stuttgart, 1948.

(16) Darré, A., "Ausgleichvorgänge bei der Schallübertragung," *Frequenz*, 6:65 (1952).

(17) Eggebrecht, H. H., "Musik als Tonsprache," *Archiv für Musikwissenschaft*, 18:73 (1961).

(18) Fant, G., "Modern Methods for Acoustic Studies of Speech," *Proceedings of the Eighth International Congress of Linguists*, Oslo, 1958.

(19) Feldtkeller, R. and Zwicker, E., *Das Ohr als Nachrichtenempfänger*, S. Hirzel, Stuttgart, 1956.

(20) Feldtkeller, R., "Hörbarkeit von Verzerrungen bei Übertragung von Instrumentenklängen," *Acustica*, 4:1 (1954).

(21) Feldtkeller, R. and Zwicker, E., "Elementarstufen der Tonhöhen- und Lautstärkeempfindung," *Akustische Beihefte* (*Acustica*), 3:97 (1953).

(22) Fletcher, H., "Pitch, Loudness and Quality of Musical Tones," *American Journal of Physics*, 14:215 (1946).

(23) Fletcher, H. and Munson, W. A., "Loudness," *Journal of the Acoustical Society of America*, 5:82 (1933).

(24) FRIELING, H., *Harmonie und Rhythmus in Natur und Kunst*, Oldenburg, Munich, 1937.

(25) GEBSER, J., *Abendländische Wandlung*, Ullstein, Frankfurt a. M., 1956.

(26) GOLD, T., *Hearing: Symposium on Information Theory*, London, 1950.

(27) GRÜTZMACHER, M. and LOTTERMOSER, W., "Tonhöhenschreiber," *Akustische Zeitschrift*, 3:183(1938).

(28) GRÜTZMACHER, M. and LOTTERMOSER, W., *Physikalische Zeitschrift*, 36:903 (1935).

(29) GRÜTZMACHER, M. and LOTTERMOSER, W., "Tonhöhenschwankungen," *Akustische Zeitschrift*, 5:1 (1940).

(30) HAAR, G., "Die Störfähigkeit quadratisch-kubischer Verzerrungen bei der Übertragung von Musik," *Frequenz*, 6:199–206 (1952).

(31) HAAS, H., "Einfluss eines Einfach-Echos auf die Hörsamkeit von Sprache," *Acustica*, 2:49 (1951).

(32) HANDSCHIN, J., *Der Toncharakter*, Atlantis, Zurich, 1948.

(33) HARTMANN, M., "Die Philosophie der Natur Nicolai Hartmanns," *Die Naturwissenschaften*, 38:468 (1951).

(34) HINDEMITH, P., *Unterweisung im Tonsatz*, B. Schott's Söhne, Mainz, 1940.

(35) HORNBOSTEL, E. M. v., "Psychologie der Gehörserscheinungen," *Handbuch der Normalen und Pathologischen Physiologie*, XI, Berlin, 1926.

(36) HUSSON, R., "L'acoustique des Salles," *Annales des Télécommunications*, 7:16 (1952).

(37) JONEN, K., *Energetik des Klavierspiels*, Mitteldeutscher Verlag, Halle, 1951.

(38) JOOS, M., *Acoustic Phonetics*, Linguistic Society of America, Baltimore, 1948.

(39) JORDAN, W., "The Building-up Process of Sound Pulses in a Room," *Proceedings of the Third ICA Congress*, Stuttgart, 1959.

(40) KATZ, D., "Beiträge zur Psychologie des Vergleichs im Gebiet des Zeitsinns," *Zeitschrift für Psychologie*, No. 4–6, p. 42 (1906).

(41) KEIDEL, W. D., "Lautheitseinfluss auf die Informationsverarbeitung beim binauralen Hören," *Pflügers Archiv*, 270:370–389 (1960).

(42) KIETZ, H., "Das räumliche Hören," *Acustica*, 3:73 (1953).

(43) KÖHLER, W., *Gestalt Psychology*, Mentor Books, New American Library, New York, 1947.

(44) KRYTER, N., "Variables Affecting Speech Communication," *Proceedings of the Second ICA Congress*, Cambridge, Mass., 1957.

(45) KÜPFMÜLLER, K., *Die Systemtheorie der elektrischen Nachrichtenübertragung*, S. Hirzel, Stuttgart, 1952.

(46) KUHL, W., "Günstigste Nachhallzeit grosser Musikstudios," *Acustica*, 4:618 (1954).

(47) KURTH, E., *Grundlagen des linearen Kontrapunkts*, Max Hesses Verlag, Berlin, 1922.

(48) LADNER, A. W., "Analysis and Synthesis of Musical Sounds," *Electronic Engineering*, 21:379 (1943).

(49) LAEY, R. E. and SAXE, R. K., "Some Aspects of Clipped Speech," *Convention Record, IRE*, p. 9 (1954).

(50) LANGENBECK, B., "Adaptation, Fowlertest und Verdeckung," *Zeitschrift für Laryngologie, Rhinologie, Otologie und ihre Grenzgebiete*, 34:778 (1955).

(51) LERCHE, E. and SCHULZE, J., "Hörermüdung und Adaptation," *Forschungs-Berichte, Nordrhein-Westfalen*, 1958.

(52) LICKLIDER, J. C. R., "Duplex Theory of Pitch Perception," *Experientia*, 7:128 (1951).

(53) LICKLIDER, J. C. R., "Correlates of the Auditory Stimulus," in S. S. Stevens, *Handbook of Experimental Psychology*, John Wiley & Sons, Inc., New York, 1951.

(54) LICKLIDER, J. C. R., "Auditory Frequency Analysis," in C. Cherry, *Information Theory, Fourth Symposium*, Butterworth & Co., Ltd., London, 1956.

(55) LOTTERMOSER, W. and BRAUNMÜHL, H. J. v., "Beitrag zur Stimmtonfrage," *Akustische Beihefte* (*Acustica*), 5:92 (1953).

(56) MARTIN, D. W. and WARD, W. D., "Subjective Evaluation of Musical Scale Temperament," *Journal of the Acoustical Society of America*, 33:582 (1961).

(57) MAYER, N., "Hörbarkeit von linearen Verzerrungen bei natürlichen Klängen," *Funk und Ton*, 5:1 (1954).

(58) MEINEL, E., "Akustische Eigenschaften klanglich hervorragender Geigen," *Akustische Zeitschrift*, 4:89 (1939).

(59) MEISTER, F. J., *Akustische Messtechnik der Gehörprüfung*, G. Braun, Karlsruhe, 1954.

(60) MEYER, E. and BUCHMANN, G., *Die Klangspektren der Musikinstrumente*, Akademie der Wissenschaften, Berlin, 1931.

(61) MEYER, E. and JORDAN, V., "Nachhallzeiten von Konzerträumen," *Elektrische Nachrichtentechnik*, 12:213 (1935).

(62) MEYER, J., *Resonanz und Einschwingvorgänge labialer Orgelpfeifen*, Technische Hochschule, Dissertation, Braunschweig, 1960.

(63) MEYER, J., "Unharmonische Komponenten im Klang der Orgelpfeifen," *Das Musikinstrument*, 11:725–730 (1962).

(64) MEYER, M. F., "Subjective Sounds," *American Journal of Psychology*, 70:646 (1957).

(65) MEYER-EPPLER, W., *Elektrische Klangerzeugung*, Dümmler, Bonn, 1949.

(66) MEYER-EPPLER, W., *Informationstheorie*, Springer-Verlag, Heidelberg, 1959.

(67) MICHEL, E., "Raumakustik," *Handbuch der Physik* (vol. VIII), Springer-Verlag, Berlin, 1927.

(68) MOLES, A., "La structure physique du signal musical," *Revue Scientifique*, Extrait 3324 (1953).

(69) MOLES, A., "Theórie de l'information," *L'Ère Atomique* (vol. VII), Kister, Geneva, 1958.

(70) MOLES, A., *Theórie de l'information et perception esthétique*, Flammarion, Paris, 1958.

(71) NICKERSON, J. E., "Untersuchung von Intonationssystemen," *Journal of the Acoustical Society of America*, 21:593 (1949).

(72) NIESE, H., "Prüfung des raumakustischen Echograd-Kriteriums," *Hochfrequenztechnik und Elektroakustik*, 66:70 (1957).

(73) NIESE, H., "Lautstärke-Empfindung von rhythmischen Geräuschen geringer Periodizität," *Hochfrequenztechnik und Elektroakustik*, 67:26 (1958).

(74) OLSON, H. F. and BELAR, H., "Electronic Music Synthesizer," *Journal of the Acoustical Society of America*, 27:595 (1955).

(75) PHILIPPOT, M., "Histoire et perspective de la réverbération," *Cahiers d'Études de Radio-Télévision*, 2:59 (1956).

(76) PIMONOW, L., "Analyse des vibrations en régime transitoire," *Annales des Télécommunications*, 113:2/11 (1958).

(77) POTTER, R. K. and PETERSEN, G. E., "The Representations of Vowels and Their Movements," *Journal of the Acoustical Society of America*, 2:528 (1948).

(78) POTTER, R. K., "Objectives for Sound Portrayal," *Journal of the Acoustical Society of America*, 21:1–5 (1949).

(79) POTTER, R. K., KOPP, G. A. and KOPP, H. G., *Visible Speech*, Dover Publications, Inc., New York, 1966.

(80) RANKE, O. F., "Hydrodynamics of the Cochlea," *Journal of the Acoustical Society of America*, 22:83 (1950).

(81) RANKE, O. F., *Physiologie des Gehörs*, Springer-Verlag, Berlin, 1953.

(82) REICHENBACH, H., *The Direction of Time*, University of California Press, Berkeley, 1956.

(83) ROSENZWEIG, M. R. and ROSENBLITH, W. A., "Responses to Auditory Stimuli," *Psychological Monographs*, 67:1 (1953).

(84) SALA, O., "Experimentelle und theoretische Grundlagen des Trautoniums," *Frequenz*, 2:315 (1948); 3:13 (1949).

(85) SCHÖNBERG, A., *Harmonielehre*, Universal Edition, Vienna, 1922.

(86) SCHOUTEN, I. F., "Theory of Residue," *Proceedings of the Academy of Amsterdam*, 43:991 (1940); "Perception of Pitch," *Philips Technical Review*, 5:286–294 (1940).

(87) SCHRÖDINGER, E., *Das Geheimnis des Lebens*, Lehnen, Munich, 1951.

(88) SCHRÖDINGER, E., *Die Natur und die Griechen*, Rowohlt, Hamburg, 1956.

(89) SEASHORE, C. E., *Psychology of Music*, McGraw-Hill Book Co., Inc., New York, 1938.

(90) SEASHORE, H., "Forms of Artistic Pitch Deviations in Singing," *Psychological Bulletin*, Princeton, 31:677 (1934).

(91) SERAPHIM, H.-P., "Unterschiedsschwelle exponentiellen Abklingens von Rauschband-Impulsen," *Acustica*, 8:280 (1958).

(92) SHANNON, C., "Mathematical Theory of Communication," *Bell System Technical Journal*, 127:379 (1948); University of Illinois Press, 1962.

(93) SKUDRZYK, E., "Die Bedeutung der Ausgleichvorgänge für Musik und Tonübertragung," *Elektrotechnik und Maschinenbau*, 67, No. 9/10 (1950).

(94) SKUDRZYK, E., "Betrachtung zum musikalischen Zusammenklang," *Acustica*, 4:18 (1954).

(95) SLOT, G., *Vom Mikrophon zum Ohr*, Philips Technische Bibliothek, Eindhoven, 1955.

(96) STERN, L. W., "Psychische Präsenzzeit," *Zeitschrift für Psychologie und Physiologie*, 13:325 (1897).

(97) STEVENS, S. S., "On the Validity of Loudness Scale," *Journal of the Acoustical Society of America*, 31:959–1003 (1959).

(98) STUMPF, C., *Die Sprachlaute*, Springer-Verlag, Berlin, 1926.

(99) STUMPF, C., *Tonpsychologie*, S. Hirzel, Leipzig, 1883 and 1890.

(100) THIELE, R., "Schallrückwürfe in Räumen," *Akustische Beihefte* (*Acustica*), 3:291 (1953).

(101) TRAUTWEIN, F., *Elektrische Musik*, Weidmann, Berlin, 1930.

(102) TRENDELENBURG, F., THIENHAUS, E. and FRANZ, E., "Klangeinsätze an der Orgel," *Akustische Zeitschrift*, 1:59 (1936); 3:7 (1938); 5:309 (1940).

(103) TRENDELENBURG, W., "Die neueren Erkenntnisse der Stimmphysiologie," *Archiv für Sprach- und Stimmphysiologie*, 6:1, No. 3/4 (1942).

(104) TÜRCK, W., "Physiologisch-akustische Kennzeichen von Ausgleichvorgängen," *Akustische Zeitschrift*, 5:129 (1940).

(105) VIERLING, O., *Das elektroakustische Klavier*, Technische Hochschule, Dissertation, Berlin, 1936.

(106) WAGNER, K. W., *Lehre von den Schwingungen und Wellen*, Diederich, Wiesbaden, 1947.

(107) WEITBRECHT, W., "Einfluss nichtlinearer Verzerrungen auf die Hörbarkeit von Verstimmungen musikalischer Intervalle," *Fernmeldetechnische Zeitschrift*, 3:336 (1950).

(108) WEIZSÄCKER, V. v., *Der Gestaltkreis*, Thieme, Stuttgart, 1950.

(109) WEIZSÄCKER, V. v., *Gestalt und Zeit* (2nd ed.), Vandenhoeck, Göttingen, 1960.

(110) WINCKEL, F., *Klangstruktur der Musik*, Verlag für Radio-Foto-Kinotechnik, Berlin, 1955.

(111) WINCKEL, F. (ed.), "Klangstruktur der Musik," article W. Lottermoser, *Untersuchungen an alten und neuen Orgeln*, Berlin, Verlag für Radio-Foto-Kinotechnik, 1955.

(112) WINCKEL, F., "Die ästhetischen Wirkungen des Vibratos," *Gravesaner Blätter*, 2:40 (1956).

(113) WINCKEL, F., "Die besten Konzertsäle der Welt," *Der Monat*, 9:75 (1957).

(114) WINCKEL, F., "Raumakustische Kriterien hervorragender Konzertsäle," *Frequenz*, 12:50 (1958).

(115) WINCKEL, F., "The Psycho-acoustical Analysis of Structure as Applied to Electronic Music," *Music Theory*, 7:194 (1963); New Haven, Yale School of Music Publication.

(116) YOUNG, R. W., "The Natural Frequencies of Musical Horns," *Acustica*, 10:91 (1960).

(117) ZWICKER, E., "Amplitudengang der nichtlinearen Verzerrungen des Ohres," *Akustische Beihefte* (*Acustica*), 5:67 (1955).

III. Supplement

(118) STEINBERG, J. C., "Positions of Stimulation in the Cochlea by Pure Tones," *Journal of the Acoustical Society of America*, 8:176 (1937).

(119) HILLER, L. A. and ISAACSON, L. M., *Experimental Music*, McGraw-Hill Book Co., Inc., New York, 1959.

(120) MOLES, A., *Les Musiques Experimentales*, Editions du Cercle d'Art Contemporain, Paris, 1960.

(121) SCHAEFFER, P., *Traité des Objects Musicaux*, Editions du Seuil, Paris, 1966.

INDEX

A CATALOGUE OF SELECTED DOVER BOOKS
IN ALL FIELDS OF INTEREST

A CATALOGUE OF SELECTED DOVER BOOKS IN ALL FIELDS OF INTEREST

RACKHAM'S COLOR ILLUSTRATIONS FOR WAGNER'S RING. Rackham's finest mature work—all 64 full-color watercolors in a faithful and lush interpretation of the *Ring*. Full-sized plates on coated stock of the paintings used by opera companies for authentic staging of Wagner. Captions aid in following complete Ring cycle. Introduction. 64 illustrations plus vignettes. 72pp. 8⅝ x 11¼. 23779-6 Pa. $6.00

CONTEMPORARY POLISH POSTERS IN FULL COLOR, edited by Joseph Czestochowski. 46 full-color examples of brilliant school of Polish graphic design, selected from world's first museum (near Warsaw) dedicated to poster art. Posters on circuses, films, plays, concerts all show cosmopolitan influences, free imagination. Introduction. 48pp. 9⅜ x 12¼. 23780-X Pa. $6.00

GRAPHIC WORKS OF EDVARD MUNCH, Edvard Munch. 90 haunting, evocative prints by first major Expressionist artist and one of the greatest graphic artists of his time: *The Scream, Anxiety, Death Chamber, The Kiss, Madonna,* etc. Introduction by Alfred Werner. 90pp. 9 x 12. 23765-6 Pa. $5.00

THE GOLDEN AGE OF THE POSTER, Hayward and Blanche Cirker. 70 extraordinary posters in full colors, from Maitres de l'Affiche, Mucha, Lautrec, Bradley, Cheret, Beardsley, many others. Total of 78pp. 9⅜ x 12¼. 22753-7 Pa. $5.95

THE NOTEBOOKS OF LEONARDO DA VINCI, edited by J. P. Richter. Extracts from manuscripts reveal great genius; on painting, sculpture, anatomy, sciences, geography, etc. Both Italian and English. 186 ms. pages reproduced, plus 500 additional drawings, including studies for *Last Supper,* Sforza monument, etc. 860pp. 7⅞ x 10¾. (Available in U.S. only) 22572-0, 22573-9 Pa., Two-vol. set $15.90

THE CODEX NUTTALL, as first edited by Zelia Nuttall. Only inexpensive edition, in full color, of a pre-Columbian Mexican (Mixtec) book. 88 color plates show kings, gods, heroes, temples, sacrifices. New explanatory, historical introduction by Arthur G. Miller. 96pp. 11⅜ x 8½. (Available in U.S. only) 23168-2 Pa. $7.95

UNE SEMAINE DE BONTÉ, A SURREALISTIC NOVEL IN COLLAGE, Max Ernst. Masterpiece created out of 19th-century periodical illustrations, explores worlds of terror and surprise. Some consider this Ernst's greatest work. 208pp. 8⅛ x 11. 23252-2 Pa. $6.00

THE ANATOMY OF THE HORSE, George Stubbs. Often considered the great masterpiece of animal anatomy. Full reproduction of 1766 edition, plus prospectus; original text and modernized text. 36 plates. Introduction by Eleanor Garvey. 121pp. 11 x 14¾. 23402-9 Pa. $6.00

BRIDGMAN'S LIFE DRAWING, George B. Bridgman. More than 500 illustrative drawings and text teach you to abstract the body into its major masses, use light and shade, proportion; as well as specific areas of anatomy, of which Bridgman is master. 192pp. 6½ x 9¼. (Available in U.S. only) 22710-3 Pa. $3.50

ART NOUVEAU DESIGNS IN COLOR, Alphonse Mucha, Maurice Verneuil, Georges Auriol. Full-color reproduction of *Combinaisons ornementales* (c. 1900) by Art Nouveau masters. Floral, animal, geometric, interlacings, swashes—borders, frames, spots—all incredibly beautiful. 60 plates, hundreds of designs. 9⅜ x 8-1/16. 22885-1 Pa. $4.00

FULL-COLOR FLORAL DESIGNS IN THE ART NOUVEAU STYLE, E. A. Seguy. 166 motifs, on 40 plates, from *Les fleurs et leurs applications decoratives* (1902): borders, circular designs, repeats, allovers, "spots." All in authentic Art Nouveau colors. 48pp. 9⅜ x 12¼. 23439-8 Pa. $5.00

A DIDEROT PICTORIAL ENCYCLOPEDIA OF TRADES AND INDUSTRY, edited by Charles C. Gillispie. 485 most interesting plates from the great French Encyclopedia of the 18th century show hundreds of working figures, artifacts, process, land and cityscapes; glassmaking, papermaking, metal extraction, construction, weaving, making furniture, clothing, wigs, dozens of other activities. Plates fully explained. 920pp. 9 x 12. 22284-5, 22285-3 Clothbd., Two-vol. set $40.00

HANDBOOK OF EARLY ADVERTISING ART, Clarence P. Hornung. Largest collection of copyright-free early and antique advertising art ever compiled. Over 6,000 illustrations, from Franklin's time to the 1890's for special effects, novelty. Valuable source, almost inexhaustible.
Pictorial Volume. Agriculture, the zodiac, animals, autos, birds, Christmas, fire engines, flowers, trees, musical instruments, ships, games and sports, much more. Arranged by subject matter and use. 237 plates. 288pp. 9 x 12. 20122-8 Clothbd. $14.50

Typographical Volume. Roman and Gothic faces ranging from 10 point to 300 point, "Barnum," German and Old English faces, script, logotypes, scrolls and flourishes, 1115 ornamental initials, 67 complete alphabets, more. 310 plates. 320pp. 9 x 12. 20123-6 Clothbd. $15.00

CALLIGRAPHY (CALLIGRAPHIA LATINA), J. G. Schwandner. High point of 18th-century ornamental calligraphy. Very ornate initials, scrolls, borders, cherubs, birds, lettered examples. 172pp. 9 x 13. 20475-8 Pa. $7.00

HOLLYWOOD GLAMOUR PORTRAITS, edited by John Kobal. 145 photos capture the stars from 1926-49, the high point in portrait photography. Gable, Harlow, Bogart, Bacall, Hedy Lamarr, Marlene Dietrich, Robert Montgomery, Marlon Brando, Veronica Lake; 94 stars in all. Full background on photographers, technical aspects, much more. Total of 160pp. 8⅜ x 11¼. 23352-9 Pa. $6.00

THE NEW YORK STAGE: FAMOUS PRODUCTIONS IN PHOTOGRAPHS, edited by Stanley Appelbaum. 148 photographs from Museum of City of New York show 142 plays, 1883-1939. *Peter Pan, The Front Page, Dead End, Our Town,* O'Neill, hundreds of actors and actresses, etc. Full indexes. 154pp. 9½ x 10. 23241-7 Pa. $6.00

DIALOGUES CONCERNING TWO NEW SCIENCES, Galileo Galilei. Encompassing 30 years of experiment and thought, these dialogues deal with geometric demonstrations of fracture of solid bodies, cohesion, leverage, speed of light and sound, pendulums, falling bodies, accelerated motion, etc. 300pp. 5⅜ x 8½. 60099-8 Pa. $4.00

THE GREAT OPERA STARS IN HISTORIC PHOTOGRAPHS, edited by James Camner. 343 portraits from the 1850s to the 1940s: Tamburini, Mario, Caliapin, Jeritza, Melchior, Melba, Patti, Pinza, Schipa, Caruso, Farrar, Steber, Gobbi, and many more—270 performers in all. Index. 199pp. 8⅜ x 11¼. 23575-0 Pa. $7.50

J. S. BACH, Albert Schweitzer. Great full-length study of Bach, life, background to music, music, by foremost modern scholar. Ernest Newman translation. 650 musical examples. Total of 928pp. 5⅜ x 8½. (Available in U.S. only) 21631-4, 21632-2 Pa., Two-vol. set $11.00

COMPLETE PIANO SONATAS, Ludwig van Beethoven. All sonatas in the fine Schenker edition, with fingering, analytical material. One of best modern editions. Total of 615pp. 9 x 12. (Available in U.S. only) 23134-8, 23135-6 Pa., Two-vol. set $15.50

KEYBOARD MUSIC, J. S. Bach. Bach-Gesellschaft edition. For harpsichord, piano, other keyboard instruments. English Suites, French Suites, Six Partitas, Goldberg Variations, Two-Part Inventions, Three-Part Sinfonias. 312pp. 8⅛ x 11. (Available in U.S. only) 22360-4 Pa. $6.95

FOUR SYMPHONIES IN FULL SCORE, Franz Schubert. Schubert's four most popular symphonies: No. 4 in C Minor ("Tragic"); No. 5 in B-flat Major; No. 8 in B Minor ("Unfinished"); No. 9 in C Major ("Great"). Breitkopf & Hartel edition. Study score. 261pp. 9⅜ x 12¼. 23681-1 Pa. $6.50

THE AUTHENTIC GILBERT & SULLIVAN SONGBOOK, W. S. Gilbert, A. S. Sullivan. Largest selection available; 92 songs, uncut, original keys, in piano rendering approved by Sullivan. Favorites and lesser-known fine numbers. Edited with plot synopses by James Spero. 3 illustrations. 399pp. 9 x 12. 23482-7 Pa. $9.95

TONE POEMS, SERIES II: TILL EULENSPIEGELS LUSTIGE STREICHE, ALSO SPRACH ZARATHUSTRA, AND EIN HELDENLEBEN, Richard Strauss. Three important orchestral works, including very popular *Till Eulenspiegel's Marry Pranks,* reproduced in full score from original editions. Study score. 315pp. 9⅜ x 12¼. (Available in U.S. only) 23755-9 Pa. $8.95

TONE POEMS, SERIES I: DON JUAN, TOD UND VERKLARUNG AND DON QUIXOTE, Richard Strauss. Three of the most often performed and recorded works in entire orchestral repertoire, reproduced in full score from original editions. Study score. 286pp. 9⅜ x 12¼. (Available in U.S. only) 23754-0 Pa. $7.50

11 LATE STRING QUARTETS, Franz Joseph Haydn. The form which Haydn defined and "brought to perfection." *(Grove's).* 11 string quartets in complete score, his last and his best. The first in a projected series of the complete Haydn string quartets. Reliable modern Eulenberg edition, otherwise difficult to obtain. 320pp. 8⅜ x 11¼. (Available in U.S. only) 23753-2 Pa. $7.50

FOURTH, FIFTH AND SIXTH SYMPHONIES IN FULL SCORE, Peter Ilyitch Tchaikovsky. Complete orchestral scores of Symphony No. 4 in F Minor, Op. 36; Symphony No. 5 in E Minor, Op. 64; Symphony No. 6 in B Minor, "Pathetique," Op. 74. Bretikopf & Hartel eds. Study score. 480pp. 9⅜ x 12¼. 23861-X Pa. $10.95

THE MARRIAGE OF FIGARO: COMPLETE SCORE, Wolfgang A. Mozart. Finest comic opera ever written. Full score, not to be confused with piano renderings. Peters edition. Study score. 448pp. 9⅜ x 12¼. (Available in U.S. only) 23751-6 Pa. $11.95

"IMAGE" ON THE ART AND EVOLUTION OF THE FILM, edited by Marshall Deutelbaum. Pioneering book brings together for first time 38 groundbreaking articles on early silent films from *Image* and 263 illustrations newly shot from rare prints in the collection of the International Museum of Photography. A landmark work. Index. 256pp. 8¼ x 11. 23777-X Pa. $8.95

AROUND-THE-WORLD COOKY BOOK, Lois Lintner Sumption and Marguerite Lintner Ashbrook. 373 cooky and frosting recipes from 28 countries (America, Austria, China, Russia, Italy, etc.) include Viennese kisses, rice wafers, London strips, lady fingers, hony, sugar spice, maple cookies, etc. Clear instructions. All tested. 38 drawings. 182pp. 5⅜ x 8. 23802-4 Pa. $2.50

THE ART NOUVEAU STYLE, edited by Roberta Waddell. 579 rare photographs, not available elsewhere, of works in jewelry, metalwork, glass, ceramics, textiles, architecture and furniture by 175 artists—Mucha, Seguy, Lalique, Tiffany, Gaudin, Hohlwein, Saarinen, and many others. 288pp. 8⅜ x 11¼. 23515-7 Pa. $6.95

"OSCAR" OF THE WALDORF'S COOKBOOK, Oscar Tschirky. Famous American chef reveals 3455 recipes that made Waldorf great; cream of French, German, American cooking, in all categories. Full instructions, easy home use. 1896 edition. 907pp. 6⅝ x 9⅜. 20790-0 Clothbd. $15.00

COOKING WITH BEER, Carole Fahy. Beer has as superb an effect on food as wine, and at fraction of cost. Over 250 recipes for appetizers, soups, main dishes, desserts, breads, etc. Index. 144pp. 5⅜ x 8½. (Available in U.S. only) 23661-7 Pa. $2.50

STEWS AND RAGOUTS, Kay Shaw Nelson. This international cookbook offers wide range of 108 recipes perfect for everyday, special occasions, meals-in-themselves, main dishes. Economical, nutritious, easy-to-prepare: goulash, Irish stew, boeuf bourguignon, etc. Index. 134pp. 5⅜ x 8½. 23662-5 Pa. $2.50

DELICIOUS MAIN COURSE DISHES, Marian Tracy. Main courses are the most important part of any meal. These 200 nutritious, economical recipes from around the world make every meal a delight. "I . . . have found it so useful in my own household,"—*N.Y. Times*. Index. 219pp. 5⅜ x 8½. 23664-1 Pa. $3.00

FIVE ACRES AND INDEPENDENCE, Maurice G. Kains. Great back-to-the-land classic explains basics of self-sufficient farming: economics, plants, crops, animals, orchards, soils, land selection, host of other necessary things. Do not confuse with skimpy faddist literature; Kains was one of America's greatest agriculturalists. 95 illustrations. 397pp. 5⅜ x 8½. 20974-1 Pa.$3.95

A PRACTICAL GUIDE FOR THE BEGINNING FARMER, Herbert Jacobs. Basic, extremely useful first book for anyone thinking about moving to the country and starting a farm. Simpler than Kains, with greater emphasis on country living in general. 246pp. 5⅜ x 8½. 23675-7 Pa. $3.50

PAPERMAKING, Dard Hunter. Definitive book on the subject by the foremost authority in the field. Chapters dealing with every aspect of history of craft in every part of the world. Over 320 illustrations. 2nd, revised and enlarged (1947) edition. 672pp. 5⅜ x 8½. 23619-6 Pa. $7.95

THE ART DECO STYLE, edited by Theodore Menten. Furniture, jewelry, metalwork, ceramics, fabrics, lighting fixtures, interior decors, exteriors, graphics from pure French sources. Best sampling around. Over 400 photographs. 183pp. 8⅜ x 11¼. 22824-X Pa. $6.00

ACKERMANN'S COSTUME PLATES, Rudolph Ackermann. Selection of 96 plates from the *Repository of Arts*, best published source of costume for English fashion during the early 19th century. 12 plates also in color. Captions, glossary and introduction by editor Stella Blum. Total of 120pp. 8⅜ x 11¼. 23690-0 Pa. $4.50

DRAWINGS OF WILLIAM BLAKE, William Blake. 92 plates from Book of Job, *Divine Comedy, Paradise Lost,* visionary heads, mythological figures, Laocoon, etc. Selection, introduction, commentary by Sir Geoffrey Keynes. 178pp. 8⅛ x 11. 22303-5 Pa. $4.00

ENGRAVINGS OF HOGARTH, William Hogarth. 101 of Hogarth's greatest works: *Rake's Progress, Harlot's Progress, Illustrations for Hudibras, Before and After, Beer Street and Gin Lane,* many more. Full commentary. 256pp. 11 x 13¾. 22479-1 Pa. $12.95

DAUMIER: 120 GREAT LITHOGRAPHS, Honore Daumier. Wide-ranging collection of lithographs by the greatest caricaturist of the 19th century. Concentrates on eternally popular series on lawyers, on married life, on liberated women, etc. Selection, introduction, and notes on plates by Charles F. Ramus. Total of 158pp. 9⅜ x 12¼. 23512-2 Pa. $6.00

DRAWINGS OF MUCHA, Alphonse Maria Mucha. Work reveals draftsman of highest caliber: studies for famous posters and paintings, renderings for book illustrations and ads, etc. 70 works, 9 in color; including 6 items not drawings. Introduction. List of illustrations. 72pp. 9⅜ x 12¼. (Available in U.S. only) 23672-2 Pa. $4.00

GIOVANNI BATTISTA PIRANESI: DRAWINGS IN THE PIERPONT MORGAN LIBRARY, Giovanni Battista Piranesi. For first time ever all of Morgan Library's collection, world's largest. 167 illustrations of rare Piranesi drawings—archeological, architectural, decorative and visionary. Essay, detailed list of drawings, chronology, captions. Edited by Felice Stampfle. 144pp. 9⅜ x 12¼. 23714-1 Pa. $7.50

NEW YORK ETCHINGS (1905-1949), John Sloan. All of important American artist's N.Y. life etchings. 67 works include some of his best art; also lively historical record—Greenwich Village, tenement scenes. Edited by Sloan's widow. Introduction and captions. 79pp. 8⅜ x 11¼. 23651-X Pa. $4.00

CHINESE PAINTING AND CALLIGRAPHY: A PICTORIAL SURVEY, Wan-go Weng. 69 fine examples from John M. Crawford's matchless private collection: landscapes, birds, flowers, human figures, etc., plus calligraphy. Every basic form included: hanging scrolls, handscrolls, album leaves, fans, etc. 109 illustrations. Introduction. Captions. 192pp. 8⅞ x 11¾. 23707-9 Pa. $7.95

DRAWINGS OF REMBRANDT, edited by Seymour Slive. Updated Lippmann, Hofstede de Groot edition, with definitive scholarly apparatus. All portraits, biblical sketches, landscapes, nudes, Oriental figures, classical studies, together with selection of work by followers. 550 illustrations. Total of 630pp. 9⅛ x 12¼. 21485-0, 21486-9 Pa., Two-vol. set $15.00

THE DISASTERS OF WAR, Francisco Goya. 83 etchings record horrors of Napoleonic wars in Spain and war in general. Reprint of 1st edition, plus 3 additional plates. Introduction by Philip Hofer. 97pp. 9⅜ x 8¼. 21872-4 Pa. $4.00

THE COMPLETE BOOK OF DOLL MAKING AND COLLECTING, Catherine Christopher. Instructions, patterns for dozens of dolls, from rag doll on up to elaborate, historically accurate figures. Mould faces, sew clothing, make doll houses, etc. Also collecting information. Many illustrations. 288pp. 6 x 9. 22066-4 Pa. $4.50

THE DAGUERREOTYPE IN AMERICA, Beaumont Newhall. Wonderful portraits, 1850's townscapes, landscapes; full text plus 104 photographs. The basic book. Enlarged 1976 edition. 272pp. 8¼ x 11¼. 23322-7 Pa. $7.95

CRAFTSMAN HOMES, Gustav Stickley. 296 architectural drawings, floor plans, and photographs illustrate 40 different kinds of "Mission-style" homes from *The Craftsman* (1901-16), voice of American style of simplicity and organic harmony. Thorough coverage of Craftsman idea in text and picture, now collector's item. 224pp. 8⅛ x 11. 23791-5 Pa. $6.00

PEWTER-WORKING: INSTRUCTIONS AND PROJECTS, Burl N. Osborn. & Gordon O. Wilber. Introduction to pewter-working for amateur craftsman. History and characteristics of pewter; tools, materials, step-by-step instructions. Photos, line drawings, diagrams. Total of 160pp. 7⅞ x 10¾. 23786-9 Pa. $3.50

THE GREAT CHICAGO FIRE, edited by David Lowe. 10 dramatic, eyewitness accounts of the 1871 disaster, including one of the aftermath and rebuilding, plus 70 contemporary photographs and illustrations of the ruins—courthouse, Palmer House, Great Central Depot, etc. Introduction by David Lowe. 87pp. 8¼ x 11. 23771-0 Pa. $4.00

SILHOUETTES: A PICTORIAL ARCHIVE OF VARIED ILLUSTRATIONS, edited by Carol Belanger Grafton. Over 600 silhouettes from the 18th to 20th centuries include profiles and full figures of men and women, children, birds and animals, groups and scenes, nature, ships, an alphabet. Dozens of uses for commercial artists and craftspeople. 144pp. 8⅜ x 11¼. 23781-8 Pa. $4.50

ANIMALS: 1,419 COPYRIGHT-FREE ILLUSTRATIONS OF MAMMALS, BIRDS, FISH, INSECTS, ETC., edited by Jim Harter. Clear wood engravings present, in extremely lifelike poses, over 1,000 species of animals. One of the most extensive copyright-free pictorial sourcebooks of its kind. Captions. Index. 284pp. 9 x 12. 23766-4 Pa. $8.95

INDIAN DESIGNS FROM ANCIENT ECUADOR, Frederick W. Shaffer. 282 original designs by pre-Columbian Indians of Ecuador (500-1500 A.D.). Designs include people, mammals, birds, reptiles, fish, plants, heads, geometric designs. Use as is or alter for advertising, textiles, leathercraft, etc. Introduction. 95pp. 8¾ x 11¼. 23764-8 Pa. $3.50

SZIGETI ON THE VIOLIN, Joseph Szigeti. Genial, loosely structured tour by premier violinist, featuring a pleasant mixture of reminiscenes, insights into great music and musicians, innumerable tips for practicing violinists. 385 musical passages. 256pp. 5⅝ x 8¼. 23763-X Pa. $4.00

PRINCIPLES OF ORCHESTRATION, Nikolay Rimsky-Korsakov. Great classical orchestrator provides fundamentals of tonal resonance, progression of parts, voice and orchestra, tutti effects, much else in major document. 330pp. of musical excerpts. 489pp. 6½ x 9¼. 21266-1 Pa. $7.50

TRISTAN UND ISOLDE, Richard Wagner. Full orchestral score with complete instrumentation. Do not confuse with piano reduction. Commentary by Felix Mottl, great Wagnerian conductor and scholar. Study score. 655pp. 8⅛ x 11. 22915-7 Pa. $13.95

REQUIEM IN FULL SCORE, Giuseppe Verdi. Immensely popular with choral groups and music lovers. Republication of edition published by C. F. Peters, Leipzig, n. d. German frontmaker in English translation. Glossary. Text in Latin. Study score. 204pp. 9⅜ x 12¼.
23682-X Pa. $6.00

COMPLETE CHAMBER MUSIC FOR STRINGS, Felix Mendelssohn. All of Mendelssohn's chamber music: Octet, 2 Quintets, 6 Quartets, and Four Pieces for String Quartet. (Nothing with piano is included). Complete works edition (1874-7). Study score. 283 pp. 9⅜ x 12¼.
23679-X Pa. $7.50

POPULAR SONGS OF NINETEENTH-CENTURY AMERICA, edited by Richard Jackson. 64 most important songs: "Old Oaken Bucket," "Arkansas Traveler," "Yellow Rose of Texas," etc. Authentic original sheet music, full introduction and commentaries. 290pp. 9 x 12. 23270-0 Pa. $7.95

COLLECTED PIANO WORKS, Scott Joplin. Edited by Vera Brodsky Lawrence. Practically all of Joplin's piano works—rags, two-steps, marches, waltzes, etc., 51 works in all. Extensive introduction by Rudi Blesh. Total of 345pp. 9 x 12. 23106-2 Pa. $14.95

BASIC PRINCIPLES OF CLASSICAL BALLET, Agrippina Vaganova. Great Russian theoretician, teacher explains methods for teaching classical ballet; incorporates best from French, Italian, Russian schools. 118 illustrations. 175pp. 5⅜ x 8½. 22036-2 Pa. $2.50

CHINESE CHARACTERS, L. Wieger. Rich analysis of 2300 characters according to traditional systems into primitives. Historical-semantic analysis to phonetics (Classical Mandarin) and radicals. 820pp. 6⅛ x 9¼.
21321-8 Pa. $10.00

EGYPTIAN LANGUAGE: EASY LESSONS IN EGYPTIAN HIEROGLYPHICS, E. A. Wallis Budge. Foremost Egyptologist offers Egyptian grammar, explanation of hieroglyphics, many reading texts, dictionary of symbols. 246pp. 5 x 7½. (Available in U.S. only)
21394-3 Clothbd. $7.50

AN ETYMOLOGICAL DICTIONARY OF MODERN ENGLISH, Ernest Weekley. Richest, fullest work, by foremost British lexicographer. Detailed word histories. Inexhaustible. Do not confuse this with *Concise Etymological Dictionary,* which is abridged. Total of 856pp. 6½ x 9¼.
21873-2, 21874-0 Pa., Two-vol. set $12.00

SECOND PIATIGORSKY CUP, edited by Isaac Kashdan. One of the greatest tournament books ever produced in the English language. All 90 games of the 1966 tournament, annotated by players, most annotated by both players. Features Petrosian, Spassky, Fischer, Larsen, six others. 228pp. 5⅜ x 8½. 23572-6 Pa. $3.50

ENCYCLOPEDIA OF CARD TRICKS, revised and edited by Jean Hugard. How to perform over 600 card tricks, devised by the world's greatest magicians: impromptus, spelling tricks, key cards, using special packs, much, much more. Additional chapter on card technique. 66 illustrations. 402pp. 5⅜ x 8½. (Available in U.S. only) 21252-1 Pa. $4.95

MAGIC: STAGE ILLUSIONS, SPECIAL EFFECTS AND TRICK PHOTOGRAPHY, Albert A. Hopkins, Henry R. Evans. One of the great classics; fullest, most authorative explanation of vanishing lady, levitations, scores of other great stage effects. Also small magic, automata, stunts. 446 illustrations. 556pp. 5⅜ x 8½. 23344-8 Pa. $6.95

THE SECRETS OF HOUDINI, J. C. Cannell. Classic study of Houdini's incredible magic, exposing closely-kept professional secrets and revealing, in general terms, the whole art of stage magic. 67 illustrations. 279pp. 5⅜ x 8½. 22913-0 Pa. $4.00

HOFFMANN'S MODERN MAGIC, Professor Hoffmann. One of the best, and best-known, magicians' manuals of the past century. Hundreds of tricks from card tricks and simple sleight of hand to elaborate illusions involving construction of complicated machinery. 332 illustrations. 563pp. 5⅜ x 8½. 23623-4 Pa. $6.00

MADAME PRUNIER'S FISH COOKERY BOOK, Mme. S. B. Prunier. More than 1000 recipes from world famous Prunier's of Paris and London, specially adapted here for American kitchen. Grilled tournedos with anchovy butter, Lobster a la Bordelaise, Prunier's prized desserts, more. Glossary. 340pp. 5⅜ x 8½. (Available in U.S. only) 22679-4 Pa. $3.00

FRENCH COUNTRY COOKING FOR AMERICANS, Louis Diat. 500 easy-to-make, authentic provincial recipes compiled by former head chef at New York's Fitz-Carlton Hotel: onion soup, lamb stew, potato pie, more. 309pp. 5⅜ x 8½. 23665-X Pa. $3.95

SAUCES, FRENCH AND FAMOUS, Louis Diat. Complete book gives over 200 specific recipes: bechamel, Bordelaise, hollandaise, Cumberland, apricot, etc. Author was one of this century's finest chefs, originator of vichyssoise and many other dishes. Index. 156pp. 5⅜ x 8. 23663-3 Pa. $2.75

TOLL HOUSE TRIED AND TRUE RECIPES, Ruth Graves Wakefield. Authentic recipes from the famous Mass. restaurant: popovers, veal and ham loaf, Toll House baked beans, chocolate cake crumb pudding, much more. Many helpful hints. Nearly 700 recipes. Index. 376pp. 5⅜ x 8½. 23560-2 Pa. $4.50

YUCATAN BEFORE AND AFTER THE CONQUEST, Diego de Landa. First English translation of basic book in Maya studies, the only significant account of Yucatan written in the early post-Conquest era. Translated by distinguished Maya scholar William Gates. Appendices, introduction, 4 maps and over 120 illustrations added by translator. 162pp. 5⅜ x 8½.
23622-6 Pa. $3.00

THE MALAY ARCHIPELAGO, Alfred R. Wallace. Spirited travel account by one of founders of modern biology. Touches on zoology, botany, ethnography, geography, and geology. 62 illustrations, maps. 515pp. 5⅜ x 8½.
20187-2 Pa. $6.95

THE DISCOVERY OF THE TOMB OF TUTANKHAMEN, Howard Carter, A. C. Mace. Accompany Carter in the thrill of discovery, as ruined passage suddenly reveals unique, untouched, fabulously rich tomb. Fascinating account, with 106 illustrations. New introduction by J. M. White. Total of 382pp. 5⅜ x 8½. (Available in U.S. only) 23500-9 Pa. $4.00

THE WORLD'S GREATEST SPEECHES, edited by Lewis Copeland and Lawrence W. Lamm. Vast collection of 278 speeches from Greeks up to present. Powerful and effective models; unique look at history. Revised to 1970. Indices. 842pp. 5⅜ x 8½. 20468-5 Pa. $8.95

THE 100 GREATEST ADVERTISEMENTS, Julian Watkins. The priceless ingredient; His master's voice; 99 44/100% pure; over 100 others. How they were written, their impact, etc. Remarkable record. 130 illustrations. 233pp. 7⅞ x 10 3/5. 20540-1 Pa. $5.95

CRUICKSHANK PRINTS FOR HAND COLORING, George Cruickshank. 18 illustrations, one side of a page, on fine-quality paper suitable for watercolors. Caricatures of people in society (c. 1820) full of trenchant wit. Very large format. 32pp. 11 x 16. 23684-6 Pa. $5.00

THIRTY-TWO COLOR POSTCARDS OF TWENTIETH-CENTURY AMERICAN ART, Whitney Museum of American Art. Reproduced in full color in postcard form are 31 art works and one shot of the museum. Calder, Hopper, Rauschenberg, others. Detachable. 16pp. 8¼ x 11.
23629-3 Pa. $3.00

MUSIC OF THE SPHERES: THE MATERIAL UNIVERSE FROM ATOM TO QUASAR SIMPLY EXPLAINED, Guy Murchie. Planets, stars, geology, atoms, radiation, relativity, quantum theory, light, antimatter, similar topics. 319 figures. 664pp. 5⅜ x 8½.
21809-0, 21810-4 Pa., Two-vol. set $11.00

EINSTEIN'S THEORY OF RELATIVITY, Max Born. Finest semi-technical account; covers Einstein, Lorentz, Minkowski, and others, with much detail, much explanation of ideas and math not readily available elsewhere on this level. For student, non-specialist. 376pp. 5⅜ x 8½.
60769-0 Pa. $4.50

THE AMERICAN SENATOR, Anthony Trollope. Little known, long unavailable Trollope novel on a grand scale. Here are humorous comment on American vs. English culture, and stunning portrayal of a heroine/villainess. Superb evocation of Victorian village life. 561pp. 5⅜ x 8½.
23801-6 Pa. $6.00

WAS IT MURDER? James Hilton. The author of *Lost Horizon* and *Goodbye, Mr. Chips* wrote one detective novel (under a pen-name) which was quickly forgotten and virtually lost, even at the height of Hilton's fame. This edition brings it back—a finely crafted public school puzzle resplendent with Hilton's stylish atmosphere. A thoroughly English thriller by the creator of Shangri-la. 252pp. 5⅜ x 8. (Available in U.S. only)
23774-5 Pa. $3.00

CENTRAL PARK: A PHOTOGRAPHIC GUIDE, Victor Laredo and Henry Hope Reed. 121 superb photographs show dramatic views of Central Park: Bethesda Fountain, Cleopatra's Needle, Sheep Meadow, the Blockhouse, plus people engaged in many park activities: ice skating, bike riding, etc. Captions by former Curator of Central Park, Henry Hope Reed, provide historical view, changes, etc. Also photos of N.Y. landmarks on park's periphery. 96pp. 8½ x 11. 23750-8 Pa. $4.50

NANTUCKET IN THE NINETEENTH CENTURY, Clay Lancaster. 180 rare photographs, stereographs, maps, drawings and floor plans recreate unique American island society. Authentic scenes of shipwreck, lighthouses, streets, homes are arranged in geographic sequence to provide walking-tour guide to old Nantucket existing today. Introduction, captions. 160pp. 8⅞ x 11¾. 23747-8 Pa. $6.95

STONE AND MAN: A PHOTOGRAPHIC EXPLORATION, Andreas Feininger. 106 photographs by *Life* photographer Feininger portray man's deep passion for stone through the ages. Stonehenge-like megaliths, fortified towns, sculpted marble and crumbling tenements show textures, beauties, fascination. 128pp. 9¼ x 10¾. 23756-7 Pa. $5.95

CIRCLES, A MATHEMATICAL VIEW, D. Pedoe. Fundamental aspects of college geometry, non-Euclidean geometry, and other branches of mathematics: representing circle by point. Poincare model, isoperimetric property, etc. Stimulating recreational reading. 66 figures. 96pp. 5⅝ x 8¼.
63698-4 Pa. $2.75

THE DISCOVERY OF NEPTUNE, Morton Grosser. Dramatic scientific history of the investigations leading up to the actual discovery of the eighth planet of our solar system. Lucid, well-researched book by well-known historian of science. 172pp. 5⅜ x 8½. 23726-5 Pa. $3.50

THE DEVIL'S DICTIONARY. Ambrose Bierce. Barbed, bitter, brilliant witticisms in the form of a dictionary. Best, most ferocious satire America has produced. 145pp. 5⅜ x 8½. 20487-1 Pa. $2.25

ART FORMS IN NATURE, Ernst Haeckel. Multitude of strangely beautiful natural forms: Radiolaria, Foraminifera, jellyfishes, fungi, turtles, bats, etc. All 100 plates of the 19th-century evolutionist's *Kunstformen der Natur* (1904). 100pp. 9⅜ x 12¼. 22987-4 Pa. $5.00

CHILDREN: A PICTORIAL ARCHIVE FROM NINETEENTH-CENTURY SOURCES, edited by Carol Belanger Grafton. 242 rare, copyright-free wood engravings for artists and designers. Widest such selection available. All illustrations in line. 119pp. 8⅜ x 11¼.
23694-3 Pa. $4.00

WOMEN: A PICTORIAL ARCHIVE FROM NINETEENTH-CENTURY SOURCES, edited by Jim Harter. 391 copyright-free wood engravings for artists and designers selected from rare periodicals. Most extensive such collection available. All illustrations in line. 128pp. 9 x 12.
23703-6 Pa. $4.50

ARABIC ART IN COLOR, Prisse d'Avennes. From the greatest ornamentalists of all time—50 plates in color, rarely seen outside the Near East, rich in suggestion and stimulus. Includes 4 plates on covers. 46pp. 9⅜ x 12¼. 23658-7 Pa. $6.00

AUTHENTIC ALGERIAN CARPET DESIGNS AND MOTIFS, edited by June Beveridge. Algerian carpets are world famous. Dozens of geometrical motifs are charted on grids, color-coded, for weavers, needleworkers, craftsmen, designers. 53 illustrations plus 4 in color. 48pp. 8¼ x 11. (Available in U.S. only) 23650-1 Pa. $1.75

DICTIONARY OF AMERICAN PORTRAITS, edited by Hayward and Blanche Cirker. 4000 important Americans, earliest times to 1905, mostly in clear line. Politicians, writers, soldiers, scientists, inventors, industrialists, Indians, Blacks, women, outlaws, etc. Identificatory information. 756pp. 9¼ x 12¾. 21823-6 Clothbd. $40.00

HOW THE OTHER HALF LIVES, Jacob A. Riis. Journalistic record of filth, degradation, upward drive in New York immigrant slums, shops, around 1900. New edition includes 100 original Riis photos, monuments of early photography. 233pp. 10 x 7⅞. 22012-5 Pa. $7.00

NEW YORK IN THE THIRTIES, Berenice Abbott. Noted photographer's fascinating study of city shows new buildings that have become famous and old sights that have disappeared forever. Insightful commentary. 97 photographs. 97pp. 11⅜ x 10. 22967-X Pa. $5.00

MEN AT WORK, Lewis W. Hine. Famous photographic studies of construction workers, railroad men, factory workers and coal miners. New supplement of 18 photos on Empire State building construction. New introduction by Jonathan L. Doherty. Total of 69 photos. 63pp. 8 x 10¾.
23475-4 Pa. $3.00

THE COMPLETE WOODCUTS OF ALBRECHT DURER, edited by Dr. W. Kurth. 346 in all: "Old Testament," "St. Jerome," "Passion," "Life of Virgin," Apocalypse," many others. Introduction by Campbell Dodgson. 285pp. 8½ x 12¼. 21097-9 Pa. $7.50

DRAWINGS OF ALBRECHT DURER, edited by Heinrich Wolfflin. 81 plates show development from youth to full style. Many favorites; many new. Introduction by Alfred Werner. 96pp. 8⅛ x 11. 22352-3 Pa. $5.00

THE HUMAN FIGURE, Albrecht Dürer. Experiments in various techniques—stereometric, progressive proportional, and others. Also life studies that rank among finest ever done. Complete reprinting of *Dresden Sketchbook*. 170 plates. 355pp. 8⅜ x 11¼. 21042-1 Pa. $7.95

OF THE JUST SHAPING OF LETTERS, Albrecht Dürer. Renaissance artist explains design of Roman majuscules by geometry, also Gothic lower and capitals. Grolier Club edition. 43pp. 7⅞ x 10¾ 21306-4 Pa. $3.00

TEN BOOKS ON ARCHITECTURE, Vitruvius. The most important book ever written on architecture. Early Roman aesthetics, technology, classical orders, site selection, all other aspects. Stands behind everything since. Morgan translation. 331pp. 5⅜ x 8½. 20645-9 Pa. $4.50

THE FOUR BOOKS OF ARCHITECTURE, Andrea Palladio. 16th-century classic responsible for Palladian movement and style. Covers classical architectural remains, Renaissance revivals, classical orders, etc. 1738 Ware English edition. Introduction by A. Placzek. 216 plates. 110pp. of text. 9½ x 12¾. 21308-0 Pa. $10.00

HORIZONS, Norman Bel Geddes. Great industrialist stage designer, "father of streamlining," on application of aesthetics to transportation, amusement, architecture, etc. 1932 prophetic account; function, theory, specific projects. 222 illustrations. 312pp. 7⅞ x 10¾. 23514-9 Pa. $6.95

FRANK LLOYD WRIGHT'S FALLINGWATER, Donald Hoffmann. Full, illustrated story of conception and building of Wright's masterwork at Bear Run, Pa. 100 photographs of site, construction, and details of completed structure. 112pp. 9¼ x 10. 23671-4 Pa. $5.50

THE ELEMENTS OF DRAWING, John Ruskin. Timeless classic by great Viltorian; starts with basic ideas, works through more difficult. Many practical exercises. 48 illustrations. Introduction by Lawrence Campbell. 228pp. 5⅜ x 8½. 22730-8 Pa. $3.75

GIST OF ART, John Sloan. Greatest modern American teacher, Art Students League, offers innumerable hints, instructions, guided comments to help you in painting. Not a formal course. 46 illustrations. Introduction by Helen Sloan. 200pp. 5⅜ x 8½. 23435-5 Pa. $4.00

THE DEPRESSION YEARS AS PHOTOGRAPHED BY ARTHUR ROTHSTEIN, Arthur Rothstein. First collection devoted entirely to the work of outstanding 1930s photographer: famous dust storm photo, ragged children, unemployed, etc. 120 photographs. Captions. 119pp. 9¼ x 10¾.
23590-4 Pa. $5.00

CAMERA WORK: A PICTORIAL GUIDE, Alfred Stieglitz. All 559 illustrations and plates from the most important periodical in the history of art photography, *Camera Work* (1903-17). Presented four to a page, reduced in size but still clear, in strict chronological order, with complete captions. Three indexes. Glossary. Bibliography. 176pp. 8⅜ x 11¼.
23591-2 Pa. $6.95

ALVIN LANGDON COBURN, PHOTOGRAPHER, Alvin L. Coburn. Revealing autobiography by one of greatest photographers of 20th century gives insider's version of Photo-Secession, plus comments on his own work. 77 photographs by Coburn. Edited by Helmut and Alison Gernsheim. 160pp. 8⅛ x 11. 23685-4 Pa. $6.00

NEW YORK IN THE FORTIES, Andreas Feininger. 162 brilliant photographs by the well-known photographer, formerly with *Life* magazine, show commuters, shoppers, Times Square at night, Harlem nightclub, Lower East Side, etc. Introduction and full captions by John von Hartz. 181pp. 9¼ x 10¾. 23585-8 Pa. $6.95

GREAT NEWS PHOTOS AND THE STORIES BEHIND THEM, John Faber. Dramatic volume of 140 great news photos, 1855 through 1976, and revealing stories behind them, with both historical and technical information. Hindenburg disaster, shooting of Oswald, nomination of Jimmy Carter, etc. 160pp. 8¼ x 11. 23667-6 Pa. $5.00

THE ART OF THE CINEMATOGRAPHER, Leonard Maltin. Survey of American cinematography history and anecdotal interviews with 5 masters—Arthur Miller, Hal Mohr, Hal Rosson, Lucien Ballard, and Conrad Hall. Very large selection of behind-the-scenes production photos. 105 photographs. Filmographies. Index. Originally *Behind the Camera.* 144pp. 8¼ x 11. 23686-2 Pa. $5.00

DESIGNS FOR THE THREE-CORNERED HAT (LE TRICORNE), Pablo Picasso. 32 fabulously rare drawings—including 31 color illustrations of costumes and accessories—for 1919 production of famous ballet. Edited by Parmenia Migel, who has written new introduction. 48pp. 9⅜ x 12¼. (Available in U.S. only) 23709-5 Pa. $5.00

NOTES OF A FILM DIRECTOR, Sergei Eisenstein. Greatest Russian filmmaker explains montage, making of *Alexander Nevsky,* aesthetics; comments on self, associates, great rivals (Chaplin), similar material. 78 illustrations. 240pp. 5⅜ x 8½. 22392-2 Pa. $4.50

AMERICAN BIRD ENGRAVINGS, Alexander Wilson et al. All 76 plates. from Wilson's *American Ornithology* (1808-14), most important ornithological work before Audubon, plus 27 plates from the supplement (1825-33) by Charles Bonaparte. Over 250 birds portrayed. 8 plates also reproduced in full color. 111pp. 9⅜ x 12½. 23195-X Pa. $6.00

CRUICKSHANK'S PHOTOGRAPHS OF BIRDS OF AMERICA, Allan D. Cruickshank. Great ornithologist, photographer presents 177 closeups, groupings, panoramas, flightings, etc., of about 150 different birds. Expanded *Wings in the Wilderness.* Introduction by Helen G. Cruickshank. 191pp. 8¼ x 11. 23497-5 Pa. $6.00

AMERICAN WILDLIFE AND PLANTS, A. C. Martin, et al. Describes food habits of more than 1000 species of mammals, birds, fish. Special treatment of important food plants. Over 300 illustrations. 500pp. 5⅜ x 8½. 20793-5 Pa. $4.95

THE PEOPLE CALLED SHAKERS, Edward D. Andrews. Lifetime of research, definitive study of Shakers: origins, beliefs, practices, dances, social organization, furniture and crafts, impact on 19th-century USA, present heritage. Indispensable to student of American history, collector. 33 illustrations. 351pp. 5⅜ x 8½. 21081-2 Pa. $4.50

OLD NEW YORK IN EARLY PHOTOGRAPHS, Mary Black. New York City as it was in 1853-1901, through 196 wonderful photographs from N.-Y. Historical Society. Great Blizzard, Lincoln's funeral procession, great buildings. 228pp. 9 x 12. 22907-6 Pa. $8.95

MR. LINCOLN'S CAMERA MAN: MATHEW BRADY, Roy Meredith. Over 300 Brady photos reproduced directly from original negatives, photos. Jackson, Webster, Grant, Lee, Carnegie, Barnum; Lincoln; Battle Smoke, Death of Rebel Sniper, Atlanta Just After Capture. Lively commentary. 368pp. 8⅜ x 11¼. 23021-X Pa. $8.95

TRAVELS OF WILLIAM BARTRAM, William Bartram. From 1773-8, Bartram explored Northern Florida, Georgia, Carolinas, and reported on wild life, plants, Indians, early settlers. Basic account for period, entertaining reading. Edited by Mark Van Doren. 13 illustrations. 141pp. 5⅜ x 8½. 20013-2 Pa. $5.00

THE GENTLEMAN AND CABINET MAKER'S DIRECTOR, Thomas Chippendale. Full reprint, 1762 style book, most influential of all time; chairs, tables, sofas, mirrors, cabinets, etc. 200 plates, plus 24 photographs of surviving pieces. 249pp. 9⅞ x 12¾. 21601-2 Pa. $7.95

AMERICAN CARRIAGES, SLEIGHS, SULKIES AND CARTS, edited by Don H. Berkebile. 168 Victorian illustrations from catalogues, trade journals, fully captioned. Useful for artists. Author is Assoc. Curator, Div. of Transportation of Smithsonian Institution. 168pp. 8½ x 9½. 23328-6 Pa. $5.00

THE SENSE OF BEAUTY, George Santayana. Masterfully written discussion of nature of beauty, materials of beauty, form, expression; art, literature, social sciences all involved. 168pp. 5⅜ x 8½. 20238-0 Pa. $3.00

ON THE IMPROVEMENT OF THE UNDERSTANDING, Benedict Spinoza. Also contains *Ethics, Correspondence,* all in excellent R. Elwes translation. Basic works on entry to philosophy, pantheism, exchange of ideas with great contemporaries. 402pp. 5⅜ x 8½. 20250-X Pa. $4.50

THE TRAGIC SENSE OF LIFE, Miguel de Unamuno. Acknowledged masterpiece of existential literature, one of most important books of 20th century. Introduction by Madariaga. 367pp. 5⅜ x 8½.
20257-7 Pa. $4.50

THE GUIDE FOR THE PERPLEXED, Moses Maimonides. Great classic of medieval Judaism attempts to reconcile revealed religion (Pentateuch, commentaries) with Aristotelian philosophy. Important historically, still relevant in problems. Unabridged Friedlander translation. Total of 473pp. 5⅜ x 8½. 20351-4 Pa. $6.00

THE I CHING (THE BOOK OF CHANGES), translated by James Legge. Complete translation of basic text plus appendices by Confucius, and Chinese commentary of most penetrating divination manual ever prepared. Indispensable to study of early Oriental civilizations, to modern inquiring reader. 448pp. 5⅜ x 8½. 21062-6 Pa. $5.00

THE EGYPTIAN BOOK OF THE DEAD, E. A. Wallis Budge. Complete reproduction of Ani's papyrus, finest ever found. Full hieroglyphic text, interlinear transliteration, word for word translation, smooth translation. Basic work, for Egyptology, for modern study of psychic matters. Total of 533pp. 6½ x 9¼. (Available in U.S. only) 21866-X Pa. $5.95

THE GODS OF THE EGYPTIANS, E. A. Wallis Budge. Never excelled for richness, fullness: all gods, goddesses, demons, mythical figures of Ancient Egypt; their legends, rites, incarnations, variations, powers, etc. Many hieroglyphic texts cited. Over 225 illustrations, plus 6 color plates. Total of 988pp. 6⅛ x 9¼. (Available in U.S. only)
22055-9, 22056-7 Pa., Two-vol. set $16.00

THE STANDARD BOOK OF QUILT MAKING AND COLLECTING, Marguerite Ickis. Full information, full-sized patterns for making 46 traditional quilts, also 150 other patterns. Quilted cloths, lame, satin quilts, etc. 483 illustrations. 273pp. 6⅞ x 9⅝. 20582-7 Pa. $4.95

CORAL GARDENS AND THEIR MAGIC, Bronsilaw Malinowski. Classic study of the methods of tilling the soil and of agricultural rites in the Trobriand Islands of Melanesia. Author is one of the most important figures in the field of modern social anthropology. 143 illustrations. Indexes. Total of 911pp. of text. 5⅝ x 8¼. (Available in U.S. only)
23597-1 Pa. $12.95

THE CURVES OF LIFE, Theodore A. Cook. Examination of shells, leaves, horns, human body, art, etc., in "*the* classic reference on how the golden ratio applies to spirals and helices in nature . . . . "—Martin Gardner. 426 illustrations. Total of 512pp. 5⅜ x 8½. 23701-X Pa. $5.95

AN ILLUSTRATED FLORA OF THE NORTHERN UNITED STATES AND CANADA, Nathaniel L. Britton, Addison Brown. Encyclopedic work covers 4666 species, ferns on up. Everything. Full botanical information, illustration for each. This earlier edition is preferred by many to more recent revisions. 1913 edition. Over 4000 illustrations, total of 2087pp. 6⅛ x 9¼. 22642-5, 22643-3, 22644-1 Pa., Three-vol. set $25.50

MANUAL OF THE GRASSES OF THE UNITED STATES, A. S. Hitchcock, U.S. Dept. of Agriculture. The basic study of American grasses, both indigenous and escapes, cultivated and wild. Over 1400 species. Full descriptions, information. Over 1100 maps, illustrations. Total of 1051pp. 5⅜ x 8½. 22717-0, 22718-9 Pa., Two-vol. set $15.00

THE CACTACEAE,, Nathaniel L. Britton, John N. Rose. Exhaustive, definitive. Every cactus in the world. Full botanical descriptions. Thorough statement of nomenclatures, habitat, detailed finding keys. The one book needed by every cactus enthusiast. Over 1275 illustrations. Total of 1080pp. 8 x 10¼. 21191-6, 21192-4 Clothbd., Two-vol. set $35.00

AMERICAN MEDICINAL PLANTS, Charles F. Millspaugh. Full descriptions, 180 plants covered: history; physical description; methods of preparation with all chemical constituents extracted; all claimed curative or adverse effects. 180 full-page plates. Classification table. 804pp. 6½ x 9¼. 23034-1 Pa. $12.95

A MODERN HERBAL, Margaret Grieve. Much the fullest, most exact, most useful compilation of herbal material. Gigantic alphabetical encyclopedia, from aconite to zedoary, gives botanical information, medical properties, folklore, economic uses, and much else. Indispensable to serious reader. 161 illustrations. 888pp. 6½ x 9¼. (Available in U.S. only) 22798-7, 22799-5 Pa., Two-vol. set $13.00

THE HERBAL or GENERAL HISTORY OF PLANTS, John Gerard. The 1633 edition revised and enlarged by Thomas Johnson. Containing almost 2850 plant descriptions and 2705 superb illustrations, Gerard's *Herbal* is a monumental work, the book all modern English herbals are derived from, the one herbal every serious enthusiast should have in its entirety. Original editions are worth perhaps $750. 1678pp. 8½ x 12¼. 23147-X Clothbd. $50.00

MANUAL OF THE TREES OF NORTH AMERICA, Charles S. Sargent. The basic survey of every native tree and tree-like shrub, 717 species in all. Extremely full descriptions, information on habitat, growth, locales, economics, etc. Necessary to every serious tree lover. Over 100 finding keys. 783 illustrations. Total of 986pp. 5⅜ x 8½. 20277-1, 20278-X Pa., Two-vol. set $11.00